Disaster Risk Reduction

Methods, Approaches and Practices

Series Editor
Rajib Shaw, Keio University, Shonan Fujisawa Campus, Fujisawa, Japan

About the Series

Disaster risk reduction is a process that leads to the safety of communities and nations. After the 2005 World Conference on Disaster Reduction, held in Kobe, Japan, the Hyogo Framework for Action (HFA) was adopted as a framework for risk reduction. The academic research and higher education in disaster risk reduction has made, and continues to make, a gradual shift from pure basic research to applied, implementation-oriented research. More emphasis is being given to multi-stakeholder collaboration and multi-disciplinary research. Emerging university networks in Asia, Europe, Africa, and the Americas have urged process-oriented research in the disaster risk reduction field. With this in mind, this new series will promote the output of action research on disaster risk reduction, which will be useful for a wide range of stakeholders including academicians, professionals, practitioners, and students and researchers in related fields. The series will focus on emerging needs in the risk reduction field, starting from climate change adaptation, urban ecosystem, coastal risk reduction, education for sustainable development, community-based practices, risk communication, and human security, among other areas. Through academic review, this series will encourage young researchers and practitioners to analyze field practices and link them to theory and policies with logic, data, and evidence. In this way, the series will emphasize evidence-based risk reduction methods, approaches, and practices.

More information about this series at https://link.springer.com/bookseries/11575

Joy Jacqueline Pereira · Mohd Khairul Zain ·
Rajib Shaw
Editors

Climate Change Adaptation in Southeast Asia

Springer

Editors
Joy Jacqueline Pereira
SEADPRI
University Kebangsaan Malaysia
Bangi, Selangor, Malaysia

Mohd Khairul Zain
SEADPRI
University Kebangsaan Malaysia
Bangi, Selangor, Malaysia

Rajib Shaw
Graduate School of Media and Governance
Keio University
Fujisawa, Kanagawa, Japan

ISSN 2196-4106 ISSN 2196-4114 (electronic)
Disaster Risk Reduction
ISBN 978-981-16-6087-0 ISBN 978-981-16-6088-7 (eBook)
https://doi.org/10.1007/978-981-16-6088-7

This Springer imprint is published by the registered company Springer Nature Singapore Pte Ltd.
The registered company address is: 152 Beach Road, #21-01/04 Gateway East, Singapore 189721, Singapore

Foreword

The ASEAN Working Group on Climate Change (AWGCC) and the ASEAN-India Green Fund supported the project on "Enhancing Local Level Climate Change Adaptation in Southeast Asia" (Phase 1) to conduct a needs assessment and scoping study of ASEAN Member States (AMS) with respect to climate change adaptation and to develop a regional climate change adaptation work programme and network. Administered by the ASEAN Secretariat, the initial phase of the ASEAN-India Green Fund Project consisted of an inception workshop and a final planning workshop, with the needs assessment and scoping study in between. I had the pleasure of jointly leading the Project with Prof. Joy Jacqueline Pereira from Universiti Kebangsaan Malaysia's Southeast Asia Disaster Prevention Research Initiative (SEADPRI-UKM).

The Inception Workshop on Enhancing Climate Change Adaptation was held in February 2015 in Bangi, Selangor in Malaysia, and the Final Planning Workshop was held in October 2015 in Manila, The Philippines. Representatives of the National Focal Points of the AWGCC participated in both the workshops. The final report on needs assessment and scoping study and establishment of the ASEAN Partner Institutions on Climate Change Adaptation (ASEANadapt) was endorsed by the AWGCC in 2016. Since then, all ASEAN Member States have ratified the Paris Agreement and communicated their Nationally Determined Contributions (NDCs) to the United Nations Framework Convention on Climate Change (UNFCCC).

I am proud to note that members of the ASEANadapt have continued to collaborate over the years to advance local level climate change adaptation in the region, with support from SEADPRI-UKM, the Asian Network on Climate Science and Technology (ANCST) and other partners. Members exchange information on good practices and communication on a regular basis and several have engaged with the Intergovernmental Panel on Climate Change (IPCC). Clearly, the capacity of researchers in the region is expanding. This book is an outcome of that continuous collaboration. It provides a current snapshot of science, policy and challenges related to climate change adaptation in the region. The findings of the stakeholder consultations that were conducted in each AMS under the ASEAN-India Green Fund Project have highlighted potential local level pilots for further funding. This is now more relevant in the post-Paris Agreement context and fast changing climate change scenario.

Foreword

The ASEAN Working Group on Climate Change (AWGCC) and the ASEAN-India Green Fund supported the project on "Enhancing Local Level Climate Change Adaptation in Southeast Asia" (Phase 1) to conduct a needs assessment and scoping study of ASEAN Member States (AMS) with respect to climate change adaptation and to develop a regional climate change adaptation work programme and network. Administered by the ASEAN Secretariat, the initial phase of the ASEAN-India Green Fund Project consisted of an inception workshop and a final planning workshop, with the needs assessment and scoping study in between. I had the pleasure of jointly leading the Project with Prof. Joy Jacqueline Pereira from Universiti Kebangsaan Malaysia's Southeast Asia Disaster Prevention Research Initiative (SEADPRI-UKM).

The Inception Workshop on Enhancing Climate Change Adaptation was held in February 2015 in Bangi, Selangor in Malaysia, and the Final Planning Workshop was held in October 2015 in Manila, The Philippines. Representatives of the National Focal Points of the AWGCC participated in both the workshops. The final report on needs assessment and scoping study and establishment of the ASEAN Partner Institutions on Climate Change Adaptation (ASEANadapt) was endorsed by the AWGCC in 2016. Since then, all ASEAN Member States have ratified the Paris Agreement and communicated their Nationally Determined Contributions (NDCs) to the United Nations Framework Convention on Climate Change (UNFCCC).

I am proud to note that members of the ASEANadapt have continued to collaborate over the years to advance local level climate change adaptation in the region, with support from SEADPRI-UKM, the Asian Network on Climate Science and Technology (ANCST) and other partners. Members exchange information on good practices and communication on a regular basis and several have engaged with the Intergovernmental Panel on Climate Change (IPCC). Clearly, the capacity of researchers in the region is expanding. This book is an outcome of that continuous collaboration. It provides a current snapshot of science, policy and challenges related to climate change adaptation in the region. The findings of the stakeholder consultations that were conducted in each AMS under the ASEAN-India Green Fund Project have highlighted potential local level pilots for further funding. This is now more relevant in the post-Paris Agreement context and fast changing climate change scenario.

In 2018, the IPCC reported that Southeast Asia is projected to be challenged by extreme climate temperatures and precipitation as well as disaster risks associated with this change. Climate change and disaster risk reduction tend to be deliberated separately in international platforms. This has contributed to the disconnect that is currently prevalent in national systems within the region. The reality is that at the local level, the issues are interlinked and connected to sustainable development. We are all aware that there are many adaptation options that can benefit and be synergistic with climate change mitigation and also serve disaster risk reduction. By focusing on the local level, as done by each AMS under the ASEAN-India Green Fund Project, there is an opportunity to implement integrated responses that link these aspects as well as other societal objectives related to sustainable development. We have laid a foundation for AMS to be a resilient and sustainable. This book is a valuable one-stop resource that documents our efforts. I would like to personally complement Prof. Joy Jacqueline Pereira for providing leadership and coordinating ASEANadapt beyond the ASEAN-India Green Fund Project.

N. H. Ravindranath
Project Leader, ASEAN-India Green Fund Project
Indian Institute of Science Bangalore
Bangalore, India

Preface

Climate change adaptation is a means to adjust to the long-term impacts of climate change, while disaster risk reduction is critical for managing the near-term effects of climate variability. Both climate change adaptation and disaster risk reduction are essential for building community resilience. Climate change adaptation can be enhanced through complementary actions across all levels, from national, state or provincial governments to local administrations. Stakeholder engagement is critical at all levels. The role of national governments is to provide strong coordination for adaptation efforts at the sub-national levels and ensure that all relevant stakeholders are engaged at each stage. National governments can facilitate adaptation actions by providing a comprehensive policy framework, legal support, adequate resources and appropriate information. Local governments, academia, civil society, non-government organizations and the private sector have to play an increasingly important role in advancing climate change adaptation and linking it to disaster risk reduction.

This book has 11 chapters where the first chapter provides an overview of climate change and key linkages to disaster risk reduction in Southeast Asia, while the remaining chapters offer a snapshot of the status of science and policy in each of the ASEAN Members States (AMS) in this context. Each chapter spotlights national policies relevant to climate change adaptation, linkages to disaster risk reduction, vulnerable ecosystems and regions, key measures, priority sectors and challenges in the respective countries. Information from government documents have been supplemented with material from scientific publications. Findings of the stakeholder consultations to delineate adaptation priorities in the respective AMS, which was conducted under the ASEAN-India Green Fund Project on "Enhancing Climate Change Adaptation in Southeast Asia" (Phase 1), are also highlighted. Pilots that were proposed for further funding in each of the AMS are also documented.

The Editors would like to record their gratitude to the ASEAN Working Group on Climate Change, ASEAN-India Green Fund Project Leader from India, Prof. N. H. Ravindranath and the Government of India for their contribution and engagement in enhancing climate change adaptation in Southeast Asia. The support of the ASEAN Secretariat, Asian Network on Climate Science and Technology (ANCST),

Asia Pacific Network for Global Change Research (APN), Asian Science, Technology and Academic Group (ASTAAG), Indian Institute of Science Bangalore (IISc. Bangalore), Universiti Kebangsaan Malaysia's Southeast Asia Disaster Prevention Research Initiative (SEADPRI-UKM) and other partners, in convening capacity building workshops is also acknowledged. The following reviewers are thanked for their assistance: Prof. Vinod K. Sharma (Indian Institute of Public Administration), Dr. Raman Letchumanan (formerly of the ASEAN Secretariat), Dr. Sugeng Triutomo (Indonesia Defense University), Ms. Antonia Yulo Loyzaga (Manila Observatory), Dr. Takako Izumi (Association of Pacific Rim Universities, APRU), Prof. Shi Peijun (UN-ISDR Asia Science Technology Academia Advisory Group, ASTAAG).

The book is written for students, researchers, academics, policymakers and development practitioners in the field of climate change adaptation and disaster risks studies. We hope that they will find the book beneficial and relevant to their work.

Kuala Lumpur, Malaysia — Joy Jacqueline Pereira
Kuala Lumpur, Malaysia — Mohd Khairul Zain
Kanagawa, Japan — Rajib Shaw

Editors and Contributors

About the Editors

Dr. Joy Jacqueline Pereira is a professor and Principal Research Fellow at Universiti Kebangsaan Malaysia's Southeast Asia Disaster Prevention Research Initiative (SEADPRI-UKM) and Fellow of the Academy of Sciences Malaysia. In 2015, she was elected as Working Group II Vice-Chair of IPCC Task Force Bureau. Prof. Pereira was Coordinating Lead Author for the Asia Chapter of the Intergovernmental Panel on Climate Change Fifth Assessment Report (IPCC-AR5) released in March 2014; Review Editor for the 2012 IPCC Special Report on Managing the Risks of Extreme Events and Disasters to Advance Climate Change Adaptation (IPCC-SREX); and Lead Author for the IPCC-AR5 Synthesis Report. She is also a Member of the United Nations Office for Disaster Risk Reduction (UNDRR) Asia Pacific Science, Technology and Academic Group (APSTAAG); and Director of the Asian Network on Climate Science and Technology (ANCST). Prof. Pereira has graduated over 40 doctoral and masters candidates and published over two hundred peer-reviewed articles.

Mohd Khairul Zain is a Senior Science Officer at Universiti Kebangsaan Malaysia's Southeast Asia Disaster Prevention Research Initiative (SEADPRI-UKM). He is currently involved in research projects related to disaster risk reduction. He also actively provides input, technical advice and support in the formulation of various government actions and negotiations at national, regional and international platforms. In 2013, Mr. Khairul became involved in the formulation of country's National Platform and Action Plan for Disaster Risk Reduction (MyDRR), which serves to integrate disaster risk reduction and climate change adaptation, engage communities and build resilience at the local level. Mr. Khairul coordinates the ASEAN Partner Institutions on Climate Change Adaptation (ASEANadapt) and founded U-INSPIRE Malaysia, a platform to empower Malaysian youth and young professionals in science, engineering, technology and innovation; to build disaster resilience at the national, regional and global levels.

Dr. Rajib Shaw is a professor in Graduate School of Media and Governance in Keio University's Shonan Fujisawa Campus (SFC). Earlier, he was the Executive Director of the Integrated Research on Disaster Risk (IRDR). He is also the Senior Fellow of Institute of Global Environmental Strategies (IGES) Japan and the Chairperson of SEEDS Asia, a Japanese NGO. Previously, he served as a Professor in the Graduate School of Global Environmental Studies of Kyoto University. He is the editor of a book series on disaster risk reduction, published by Springer, and is the Chair of UN ISDR's Global Science Technology Advisory Group (Global STAG) and a Co-chair of UN ISDR's Asia Pacific Science Technology Academic Advisory Group (APSTAAG). Prof. Shaw is also Coordinating Lead Author for the Asia Chapter of the Intergovernmental Panel on Climate Change Sixth Assessment Report (IPCC-AR6). He has published more than 40 books and over 300 academic papers and book chapters.

Contributors

Thannob Aribarg Climate Change and Disaster Center, Rangsit University, Rangsit, Thailand

Andini Desita Ekaputri National Research and Innovation Agency (BRIN) Indonesia, Jakarta, Indonesia

Ankit Joshi Independent Advisor, Asian Disaster Preparedness Center, Bangkok, Thailand

Luh Kitty Katherina National Research and Innovation Agency (BRIN) Indonesia, Jakarta, Indonesia

Vilakone Maniphousay Department of Climate Change (DCC), Ministry of Natural Resources and Environment, Vientiane, Lao PDR

Sothun Nop Faculty of Development Studies, Royal University of Phnom Penh, Phnom Penh, Cambodia;
School of Science, The University of New South Wales, Canberra, Australia

Mitsuko Shikada Otsuyama SEEDS Asia, Kobe, Japan

Joy Jacqueline Pereira Universiti Kebangsaan Malaysia, Bangi, Malaysia

Juan M. Pulhin University of the Philippines, Los Baños, Philippines

Laksmi Rachmawati National Research and Innovation Agency (BRIN) Indonesia, Jakarta, Indonesia

Heru Santoso National Research and Innovation Agency (BRIN) Indonesia, Jakarta, Indonesia

About This Book

This book provides an overview of climate change adaptation in ten ASEAN Members States (AMS). The chapters have been prepared by members of the ASEAN Partner Institutions on Climate Change Adaptation (ASEANadapt). The ASEANadapt resulted from the ASEAN-India Green Fund Project on "Enhancing Climate Change Adaptation in Southeast Asia" (Phase 1), an initiative under the aegis of the ASEAN Working Group on Climate Change (AWGCC) managed by the ASEAN Secretariat.

The book draws on government documents and scientific publications to spotlight national policies relevant to climate change adaptation, linkages to disaster risk reduction, vulnerable ecosystems and regions, key measures, priority sectors and challenges. Adaptation priorities were delineated through stakeholder consultations in the respective AMS and potential pilots have been highlighted for funding.

Covering aspects of climate change adaptation and disaster risks, this book is a valuable one-stop resource on the status of science and policy in Southeast Asia for students, researchers, academics, policymakers and development practitioners, who will be able to apply the knowledge for informed decision-making.

About This Book

Andreas Schaffer Independent Advisor, Asian Disaster Preparedness Center, Bangkok, Thailand

Rajib Shaw Keio University, Fujisawa, Japan

Rahayu Sukmaria Sukri Universiti Brunei Darussalam, Bandar Seri Begawan, Brunei

Seree Supratid Climate Change and Disaster Center, Rangsit University, Rangsit, Thailand

Maricel A. Tapia University of the Philippines, Los Baños, Philippines

Trong Dinh Tran Vietnam Institute of Meteorology, Hydrology and Climate Change, Hanoi, Vietnam

Thang Van Nguyen Vietnam Institute of Meteorology, Hydrology and Climate Change, Hanoi, Vietnam

Dennis Wah Universiti Brunei Darussalam, Bandar Seri Begawan, Brunei

Mohd Khairul Zain Universiti Kebangsaan Malaysia, Bangi, Malaysia

Contents

1 **Southeast Asia: An Outlook on Climate Change** 1
Joy Jacqueline Pereira and Rajib Shaw

2 **Climate Change Adaptation in Brunei Darussalam** 25
Rahayu Sukmaria Sukri and Dennis Wah

3 **Climate Change Adaptation in Cambodia** 43
Sothun Nop

4 **Climate Change Adaptation in Indonesia** 57
Laksmi Rachmawati, Andini Desita Ekaputri, Luh Kitty Katherina, and Heru Santoso

5 **Climate Change Adaptation in Lao PDR** 77
Vilakone Maniphousay

6 **Climate Change Adaptation in Malaysia** 103
Joy Jacqueline Pereira and Mohd Khairul Zain

7 **Climate Change Adaptation in Myanmar** 117
Mohd Khairul Zain, Mitsuko Shikada Otsuyama, and Rajib Shaw

8 **Climate Change Adaptation in the Philippines** 129
Juan M. Pulhin and Maricel A. Tapia

9 **Climate Change Adaptation in Singapore** 175
Ankit Joshi and Andreas Schaffer

10 **Climate Change Adaptation in Thailand** 197
Seree Supratid and Thannob Aribarg

11 **Climate Change Adaptation in Vietnam** 217
Trong Dinh Tran and Thang Van Nguyen

Chapter 1
Southeast Asia: An Outlook on Climate Change

Joy Jacqueline Pereira and Rajib Shaw

Abstract Climate change and disaster risks are mainstreamed in the three major pillars of the Association of Southeast Asian Nations (ASEAN) and complement the UN 2030 Agenda for Sustainable Development. All ASEAN Member States (AMS) have ratified the Paris Agreement. The projected climate change impacts for Southeast Asia are less severe for global warming of 1.5 °C compared with 2 °C or more. Overall, there is only a medium level of readiness in handling the projected impacts for global warming of 1.5 °C in the region, except for prioritized adaptation measures to address reduction in rice yields in several AMS. Transitioning to a 1.5 °C world requires astute adaptation, linking climate change and disaster risk reduction, amplified engagement with targeted stakeholders, and smart partnerships to hasten the process of developing truly effective climate change adaptation plans in AMS. Greater ambition is also required to create a healthy environment in AMS. Potential strategic initiatives that could benefit AMS will require mobilization of resources and technical support. All nations of the world have a critical role to play in limiting global warming to 1.5 °C.

Keywords Climate change · Vulnerability · Disaster risk · ASEAN · Southeast Asia

1.1 Introduction

Southeast Asia has ten countries that are members of the Association of Southeast Asian Nations (ASEAN), which was established in 1967 with a focus on promoting stability and economic growth in the region. The ASEAN Member States (AMS) are Brunei Darussalam, Cambodia, Indonesia, Lao PDR, Malaysia, Myanmar, the Philippines, Singapore, Thailand, and Vietnam (Fig. 1.1). With a total area of 4.4

J. J. Pereira (✉)
Uuniversiti Kebangsaan Malaysia, 43600 Bangi, Malaysia
e-mail: joy@ukm.edu.my

R. Shaw
Keio University, Fujisawa, Japan

J. J. Pereira et al. (eds.), *Climate Change Adaptation in Southeast Asia*, Disaster Risk Reduction, https://doi.org/10.1007/978-981-16-6088-7_1

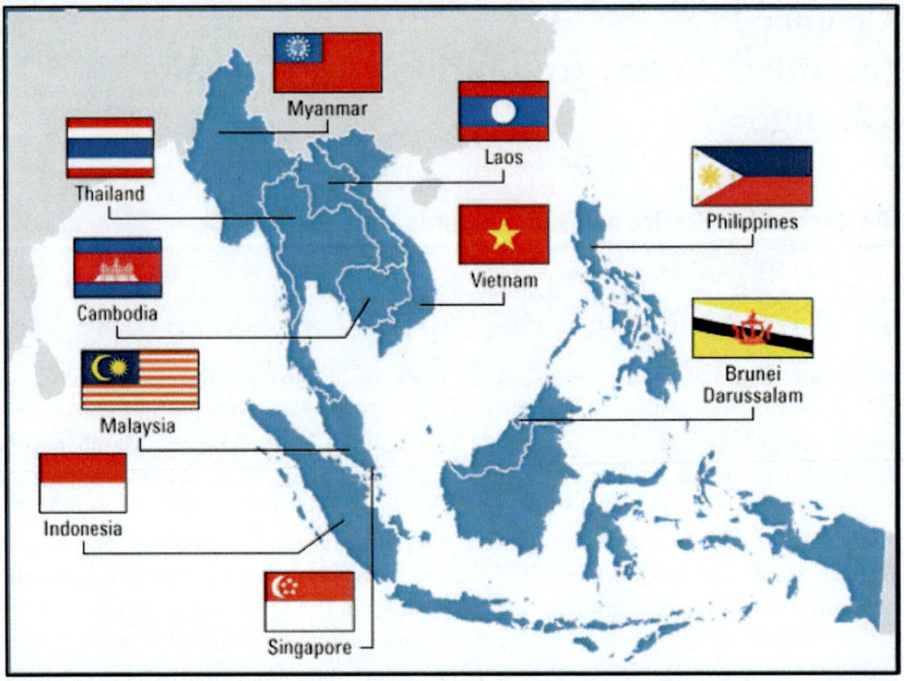

Fig. 1.1 ASEAN member states comprise 10 countries in Southeast Asia

million square kilometers, and a population of over 622 million representing about 8.5% of the global populace, the AMS currently have the third largest labor force in the world (WPR 2020). The collective economic growth of ASEAN has increased remarkably since 1999, from USD 577 billion in 1999 to USD 2.5 trillion in 2016, with an average growth of 5.3% in 2018 (OECD 2019). Among the AMS, Singapore, Brunei and Malaysia are categorized as very high on the human development index, while Thailand, the Philippines, and Indonesia are in the high category, with Cambodia, Laos PDR, Myanmar, and Vietnam falling within the medium category (UNDP 2019). Economic growth has been slower than forecasted due to global geopolitical challenges and the COVID-19 pandemic.

Between 1970 and 2010, Southeast Asia was one of the three regions of the world (in addition to Europe and China) that experienced the largest conversion of agricultural lands to urban land uses (Güneralp et al. 2020). Urbanization and industrialization have driven economic growth that benefitted the region but also led to unbridled energy demand. Energy efficiency has not improved in many AMS due to weak regulations and lack of market-based pricing (Shi 2016). The combined sectors of manufacturing and transportation have been identified as the primary energy consumer in Indonesia, Malaysia, the Philippines, Singapore, Thailand, and Vietnam (Setyawan 2020). Economic growth has also contributed to environmental and health impacts associated with air pollutants such as nitrogen oxides, sulfur dioxide and particulate matter in Southeast Asia (Fong et al. 2020). The emission of carbon dioxide and particulate matter (PM2.5) has recently been identified as major risk factors for lung cancer in the region (Farhad and Farzad 2020). Increasing use of renewable energy and higher investment in healthcare expenditure per capita have been linked to decreased prevalence of lung cancer.

Climate change and natural disasters are recognized as a threat to advancement in Southeast Asia, disproportionally affecting the poor and marginalized (ADRMI 2010; The ASEAN Post 2019). Climate-related hazards in the region include floods, storms, tropical cyclones, droughts, and extreme temperatures; 62% of the population is exposed to drought and about 24% to floods (ESCAP 2020). The landscape for risk is contributed by droughts (60%), floods (22.5%), tropical cyclones (14.3%), earthquakes (4.1%), and tsunami (0.2%), based on impacted population, geographical area, and economic losses. Among the AMS, the average annual loss as a percentage of GDP associated with intensive risk, extensive risk, indirect losses, and agriculture drought, is highest in Lao PDR (8.7%), followed by Cambodia (8%), Myanmar (6.2%), and Vietnam (5.6%) compared with the other countries in the region, with Singapore having the lowest value at 0.5% (ESCAP 2020).

The outlook on climate change in Southeast Asia commences with an overview of institutions and arrangements in ASEAN that support the region in addressing climate change, including the status of participation of AMS in global climate agreements, specifically the United Nations Framework Convention on Climate Change (UNFCCC). This is followed by a short description of the prognosis for 1.5 °C global warming and beyond in the region, based on the latest available science of the IPCC. A cursory evaluation of the readiness of the AMS in tackling emerging issues, drawing on information from the national communication of the AMS under the UNFCCC is then presented. This provides the context for discussing key challenges and opportunities for the AMS in transitioning to a 1.5 °C world, drawing on findings from the subsequent chapters and other sources. The chapter concludes with a highlight of potential initiatives that could be supported in the region to benefit AMS.

1.2 Regional Arrangements

1.2.1 ASEAN and Climate Change

ASEAN was established in 1967 and the ASEAN Charter, adopted in 2007, is a legally binding agreement recognized by the United Nations; it frames the institutional arrangements and codifies the norms, rules, and values, serving as the foundation of the ASEAN Community (ASEAN 2020). The Declaration on ASEAN 2025: Forging Ahead Together was signed by all the AMS in Kuala Lumpur to complement the UN 2030 Agenda for Sustainable Development. The Declaration on ASEAN 2025 comprises three major pillars, the Political-Security Community

Blueprint 2025, ASEAN Economic Community Blueprint 2025 and ASEAN Socio-Cultural Community Blueprint (ASCC) 2025; the latter succeeds the ASEAN Socio-Cultural Community Blueprint (ASCC) 2009–2015. Climate change and disaster risks are mainstreamed in all the three pillars of the Declaration on ASEAN 2025. Highlights are as follows:

- The Political-Security Community Blueprint 2025 emphasizes disaster management and emergency response in the region. More importantly, the Blueprint also includes the element to "Ensure that disaster risk reduction is integrated into ASEAN strategies on disaster management and emergency response"; this supports implementation of the Sendai Framework on Disaster Risk Reduction in the region.
- The ASEAN Economic Community Blueprint 2025 is envisioned to "Create a more dynamic and resilient ASEAN, capable of responding and adjusting to emerging challenges through robust national and regional mechanisms that address food and energy security issues, natural disasters, economic shocks, and other emerging trade-related issues as well as global mega trends," among other matters. Aspects emphasized include science-based support for green technology and energy, energy efficiency and renewable energy, reduction in greenhouse gas emission in agriculture, increased resilience to climate change, natural disasters and other shocks, and increased resilience to climate change in the context of food security.
- The ASEAN Socio-Cultural Community (ASCC) Blueprint 2025 aims to deliver and fully realize human development, resiliency, and sustainable development within the region. A key objective is to deliver "a resilient community with enhanced capacity and capability to adapt and respond to social and economic vulnerabilities; disasters, climate change as well as emerging threats, and challenges."

The ASEAN Working Group on Climate Change (AWGCC) was established in 2009, with the obligation to oversee the implementation of climate change actions incorporated in the ASCC Blueprint 2009–2015. The Action Plan on Joint Response to Climate Change was developed in 2012, paving the way for more nuanced implementation of the ASCC Blueprint in conjunction with dialogue partners and other organizations; the ASCC Blueprint 2009–2015 was succeeded by the ASCC Blueprint 2025. The AWGCC also supports the issuance of periodic joint statements of the ASEAN Heads of State and Government to express their common positions in addressing climate change. For example, the 2014 Statement of ASEAN Heads of State and Government agreed that the AMS would improve collective capacity to address climate change and strengthen rapid response through existing mechanisms under the ASEAN Agreement on Disaster Management and Emergency Response (AADMER). The 2019 Statement of ASEAN Heads of State and Government agreed that the AMS would have collective targets on energy efficiency of vehicles and renewable energy.

Climate change aspects are explicit in the AWGCC, which is under the ASEAN Ministerial Meeting on the Environment (AMME), ASEAN Senior Officials on the

Environment (ASOEN). The linkages to disaster risks and energy are also addressed to varying extents under the ASEAN Ministerial Meeting on Disaster Management (AMMDM), ASEAN Committee on Disaster Management (ACDM) and ASEAN Coordinating Centre for Humanitarian Assistance on Disaster Management (AHA), as well as the ASEAN Ministers on Energy Meeting (AMEM), Senior Officials Meeting on Energy (SOME), ASEAN Transport Ministers Meeting (ATM), Senior Transport Officials Meeting (STOM), and ASEAN Centre for Energy (ACE). In addition, there are several ASEAN Sectoral Ministerial Bodies in all the three major pillars where climate change adaptation, mitigation, and disaster risks are implicit (ASEAN 2020). However, the extent to which these issues are addressed vary. Such bodies include the ASEAN Regional Forum, ASEAN Regional Forum Senior Officials Meeting (ARF SOM), ASEAN Ministers Meeting on Agriculture and Forestry (AMAF), Senior Officials Meeting of the ASEAN Ministers Meeting on Agriculture and Forestry (SOM-AMAF), ASEAN Senior Officials on Forestry (ASOF), Conference of the Parties to the ASEAN Agreement on Transboundary Haze Pollution (COP), ASEAN Centre for Biodiversity (ACB), ASEAN Specialized Meteorological Centre (ASMC), and the ASEAN University Network (AUN), among others.

The Declaration on ASEAN 2025 requires translation into national systems and the three major blueprints are supported by sector-specific master plans. All AMS have ratified the Paris Agreement and communicated their Nationally Determined Contributions (NDCs) as of 2019 (Fig. 1.2). All AMS have also submitted their Initial National Communications (NC1) and Second National Communications (NC2) to

Brunei Darussalam						NC1 NDC1 2016	NC2 2017		
Cambodia		NC1 2002	NAPA 2007			NC2 2016	NDC1 2017		
Indonesia	NC1 1999			NC2 2011		NDC1 2016		NC3 2018	
Lao PDR	NC1 2000		NAPA 2009		NC2 2013	NDC1 2016			
Malaysia	NC1 2000			NC2 2011		NDC1 2016		NC3 2018	
Myanmar				NC1 2012	NAPA 2013		NDC1 2017		
Philippines	NC1 2000				NC2 2014		INDC 2017		
Singapore	NC1 2000			NC2 2011	NC3 2014	NDC1 2016		NC4 2018	NDC1 Update 2020
Thailand	NC1 2000			NC2 2011		NDC1 2016		NC3 2018	
Vietnam		NC1 2003		NC2 2010		NDC1 2016			NC3 2019

Fig. 1.2 Submission of various reports on climate change by ASEAN member states as reflected in the UNFCCC Registry. *Source* https://unfccc.int/ accessed on 19 July 2020

the UNFCCC, except for Myanmar which is still in the process of preparing their NC2. Indonesia, Thailand, and Vietnam have already submitted their Third National Communications (NC3), while all the other AMS are working to communicate their NC3; Singapore has already submitted their fourth National Communication (NC4). The Paris Agreement encourages all parties to engage in the formulation and implementation of National Adaptation Plans (NAPs). Initiatives have commenced in some AMS to address this matter. A National Adaptation Program of Action (NAPA) is a type of plan similar to NAPs submitted by the least developed countries to the UNFCCC. Cambodia, Lao PDR, and Myanmar, have submitted their NAPA to identify urgent and important actions to adapt to climate change.

1.2.2 ASEAN Collaborations on Climate Change

The ASEAN Regional Forum (ARF) is a broad-based political and security cooperation platform for creating constructive dialogue among AMS and partners such as the European Union (EU), Australia, Canada, China, India, Japan, the Republic of Korea, New Zealand, and the United States. The ARF has facilitated collaboration on issues related to disaster management and emergency response as well as climate change, among others, which then come under the purview of the AWGCC, and also other relevant working groups on the environment, agriculture and forestry, energy and transport, and science and technology sectors. Some key regional initiatives on climate change that laid the foundation for regional collaboration include the following:

- Rehabilitation and Sustainable Use of Peatland Forests in Southeast Asia (2009–2013): The focus was on measures to prevent peatland fires, the major source of haze in the region;
- Biodiversity and Climate Change (2010–2015): The emphasis was on developing and implementing strategies and instruments related to biodiversity and climate change;
- Workshop and Exchange on Climate Resilient Cities (2010): The event was held in Jakarta (18–19 January 2010) to bring together officials from ASEAN cities and national governments to exchange best practices, lessons learned, and experiences in addressing climate change impacts;
- Workshops on Risks and Impacts from Extreme Events of Floods and Droughts in ASEAN (2010): Events were held in Indonesia (9–10 June 2010) and Thailand (22–23 September 2010), respectively, to assess the capacity of AMS on flood and drought management, review preparedness to engage in risk mitigation and adaptation planning, and exchange best practices on solutions;
- ASEAN Environmental Education Plan (2008–2012): Two events were convened under this initiative in 2010: (i) the ASEAN Plus Three Youth Environment Forum: Creating a Climate for Change, held in Brunei Darussalam, to generate interest and awareness of youths on climate change issues and seek their pledge to play

their part in safeguarding the environment and (ii) the annual ASEAN Plus Three Leadership Program on Sustainable Production and Consumption: to equip business and industry leaders with knowledge, skills, and tools to develop strategies for sustainable development;

- ASEAN Environmentally Sustainable Development Film Festival on Climate Change (2011): The event was held in Phnom Penh, Cambodia to showcase unique, creative, indigenous, and impactful initiatives taken by various stakeholders in ASEAN countries to inspire and promote awareness of multistakeholder participation in addressing climate change;
- The Yogyakarta City Greenhouse Gases (GHG) Emissions and HEAT + (2012): A collaboration with the International Council for Local Environmental Initiatives (ICLEI)—Local Governments for Sustainability, the ASEAN-US Technical Assistance and Training Facility (ASEAN-US TATF). The initiative involved training and the launch of the Yogyakarta City Greenhouse Gases (GHG) Emission Inventory Report. Yogyakarta was the ASEAN pilot city to demonstrate a systematic and standardized methodology to measure and monitor city-wide carbon emissions. The pilot served as a model for other ASEAN cities in developing effective strategies for low carbon economic growth and climate resiliency.

1.2.3 ASEAN-India Collaboration

In 2012, the Expert Meeting on Regional Program of Climate Change was held in Bangalore, India, to mark the commencement of initiatives supported by the ASEAN-India Green Fund, under the purview of the AWGCC, with administrative support from the ASEAN Secretariat. The purpose was to exchange information, develop a framework for collaboration, and discuss ways forward to address climate change adaptation and mitigation. This was followed by the initiation of several ASEAN-India Green Fund projects. An initial project was on "Enhancing Local Level Climate Change Adaptation in Southeast Asia," administered by the ASEAN Secretariat, which consisted of an inception workshop and a final planning workshop, with the needs assessment and scoping study in between. The Inception Workshop on Enhancing Climate Change Adaptation was held in February 2015 in Bangi, Selangor in Malaysia, and the Final Planning Workshop and regional engagement was held in October 2015 in Manila, the Philippines. Representatives of the National Focal Points of the AWGCC participated in both the workshops. The final report on needs assessment and scoping study as well as establishment of the ASEAN Partner Institutions on Climate Change Adaptation (ASEANadapt) was endorsed by the AWGCC in 2016.

The project documented a broad range of national needs and priorities across the AMS. Generally, climate change is mainstreamed at the national level but its embedment in governance processes at the local level is relatively limited. In conducting the needs assessment, stakeholder engagement in many AMS was limited, focusing primarily on the government sector. The regional engagement of 2015 involved representatives from entities such as the Asian Development Bank (ADB), Asia Pacific Network for Global Change Research (APN), Asian Science, Technology and Academic Group (ASTAAG), Asian Network on Climate Science and Technology (ANCST), the University of Cambridge, the University of Keio, Universiti Kebangsaan Malaysia, and the University of the Philippines. The aspects noted include the following:

- Regional requirements for climate change adaptation include both conventional research and action-oriented projects, capacity building, advocacy of policy and decision-makers, public awareness about climate change impacts, climate risk communication, early warning for climate extremes and best practices case studies, and promotion of public–private participation;
- Platforms for sharing and exchange of information relevant to climate change would advance the climate change adaptation agenda in the region, specifically if it served to maintain ties among AMS researchers and practitioners, who have a lot of experience and success stories that could be shared;
- A flagship regional initiative on climate change assessment (similar to that conducted by the Intergovernmental Panel on Climate Change, IPCC) led and implemented by ASEAN researchers on a periodic basis and administered by the ASEAN Secretariat would serve the governments of AMS and the international community in making evidence-based decision-making for the region.
- Funding should give priority to the development of local adaptation plans as communities are the most vulnerable, with researchers in ASEAN being mobilized to support the development of local adaptation action plans, and embedment into local governance systems via selected pilots so that they can then be replicated by the stakeholders.
- The participation of local governments in climate change adaptation is important and necessary to effectively tackle the impacts of climate change as they are the ultimate implementers. However, coordination at the national level is critical to make this happen. Emphasis should be given to build capacity and knowledge management skills of local government units.

The project also concluded that within the national boundaries of the AMS, there are variations in local conditions and this is where the problems need to be tackled. Thus, the emphasis should be on specific areas such as a landscape or an ecosystem, or a watershed, which has clusters of settlements, where both the aspects of adaptation and mitigation can be considered. Short-term risks can be handled via immediate disaster risk management programs, where climate modeling outputs are not required. Long term planning and resilience development program can draw insights from climate modeling. This can be handled using the same context and area-specific approach. Notwithstanding, there are common tools, techniques, methods, and approaches that are applicable across the ten AMS that require a self-sustaining long-term collaborative effort to enable sharing and exchange of information relevant to climate change among the AMS researchers and practitioners. This was the basis for establishing a virtual network of ASEAN Partner Institutions on Climate Change

Adaptation (ASEANadapt), to continue with exchange of information on good practices and communication beyond the project and advance the climate change adaptation agenda in the region (https://aseanadapt.org). ASEANadapt is hosted and coordinated by Universiti Kebangsaan Malaysia's Southeast Asia Disaster Prevention Initiative (SEADPRI-UKM).

1.3 Regional Climate Change Risks and Key Issues

1.3.1 Implications of 1.5 °C Global Warming

The cycle of the IPCC Sixth Assessment Report (AR6) includes the preparation of three Special Reports that precedes the main Assessment Report. The Special Report on the impacts of Global Warming of 1.5 °C above pre-industrial levels (IPCC SR1.5 °C) was released in 2018. The report was prepared in response to the invitation of the 21st Conference of Parties of the UNFCCC. The Special Reports on Climate Change and Land (IPCC SRCCL) and Ocean and Cryosphere in a Changing Climate (IPCC SROCC) were released in 2019. The IPCC Special Reports are based on the assessment of the available scientific, technical, and socioeconomic literature relevant to global warming, and among the key findings include several projected impacts that are specific to the tropics and relevant to Southeast Asia (Table 1.1).

The IPCC SR1.5 °C confirmed that humanity and its ecosystems are already impacted by warming of 1 °C since pre-industrial times. Current development is expected to lead humanity to global warming of 1.5 °C between 2030 and 2052 but this track can be altered. Apathy is expected to result in a range of impacts affecting some tens to hundred million people in the world, where the intensity is geographically differentiated. As tropical Southeast Asia moves toward 2 °C, the impacts of 1.5 °C include higher levels of precipitation associated with tropical cyclones; greater net reductions in yields of maize, rice, and other cereal crops as well as CO_2-dependent nutritional quality of rice; increased number of hot days; the largest impacts on economic growth due to climate change. Many of the projected impacts could be avoided by limiting global warming to 1.5 °C compared with 2 °C or more.

The IPCC SRCCL states that lives and livelihoods in cyclone-prone areas of the region are expected to be exposed to land degradation resulting from the combination of sea level rise and more intense cyclones. While regional warming is expected to reduce in areas where increased rainfall is projected due to increased growth of vegetation and wetter soil conditions, areas with drier soil conditions will experience more severe heatwaves. Afforestation and reforestation in the region is expected to contribute to cooling and lower the scale of heat-related events due to greater evapotranspiration. Warming is projected to result in the emergence of unprecedented climatic conditions by the mid to late twenty-first century under medium and high GHG emissions scenarios by mid to late twenty-first century.

Table 1.1 A cursory view of readiness* of AMS in handling projected impacts for Southeast Asia based on adaption measures reported in their respective National Communications (NC)

Region-Specific Projected Impact (extracted from three IPCC Special Reports**)	ASEAN MEMBER STATES (AMS)									
	BD	Cam	Ind	Lao	Msia	Myn	Phi	Sing	Tha	Viet
Heavy precipitation associated with tropical cyclones (medium confidence) [1]	?	✋	?	✋	?	✋	✋	?	✋	✋
Land degradation from combined sea level rise and cyclones jeopardise lives and livelihoods in cyclone prone areas (very high confidence) [2]	✋	✋	✋	NR	✋	✋	✋	✋	✋	✋
Increase in the number of hot days in most land regions, with highest increases in the tropics (high confidence) [1]	✋	✋	✋	✋	✋	✋	✋	✋	✋	✋
Emergence of unprecedented climatic conditions by the mid to late 21st century (medium confidence) [2]	✋	✋	✋	✋	✋	✋	✋	✋	✋	✋
Net reductions in yields of maize, rice, wheat, and potentially other cereal crops (IPCC SR 1.5°C) (high confidence) [1]	😊	✋	😊	✋	😊	✋	😊	✋	😊	😊
Net reductions in the CO_2-dependent nutritional quality of rice and wheat (high confidence) [1]	✋	✋	✋	✋	✋	✋	✋	✋	✋	✋
Region expected to experience the largest impacts on economic growth due to climate change (medium confidence) [1]	✋	✋	✋	✋	✋	✋	✋	✋	✋	✋
Increase in frequency, duration, extent and intensity of marine heatwaves by 20 times (RCP2.6, medium confidence) [3]	✋	✋	✋	NR	✋	✋	✋	✋	✋	✋
Significant changes in wave heights and coastal tidal amplitudes and patterns (high confidence) [3]	✋	✋	✋	NR	✋	✋	✋	✋	✋	✋
Decrease in biomass of marine animal communities, production, and fisheries catch potential in all scenarios (high confidence) [3]	✋	💣	✋	NR	✋	✋	✋	✋	✋	✋
Decline in ocean net primary production by 7–16% (very likely range) [3]	✋	💣	✋	NR	✋	✋	✋	✋	✋	✋
Widespread challenges to fisheries governance in regional hotspots (medium confidence) [3]	?	?	?	NR	?	?	?	?	?	?

😊	High Level of Readiness: Region-specific projected impact is stated in the NC and adaptation measures are a priority.	💣	Low Level of Readiness: Region-specific projected impact is not stated in the NC nor is the associated sector a priority for adaptation.
✋	Medium Level of Readiness: Region-specific projected impact is not stated in the NC but the associated sector is a priority for adaptation.	?	Low Level of Readiness: Region-specific projected impact has to be investigated in the AMS.
		NR	Not Relevant: Region-specific projected impact is not applicable to the AMS.

*Readiness in this context refers to the adaptation planning measures where actual implementation for many AMS depends on the availability of resources and technical support

**The ocean projections are under RCP8.5 scenarios unless stated otherwise. In the RCP2.6 scenario, surface warming stays below a 2 °C threshold at the end of the twenty-first century, and for RCP8.5, the range is 3.5–4.5 °C in 2100

Sources for region-specific projected impact:[a]IPCC SR1.5 °C,[b]IPCC SRCCL, and [c]IPCC SROCC

Sources for adaptation priorities: NC2 Brunei Darussalam (2017), NC2 Cambodia (2015), NC3 Indonesia (2017), NC2 Lao PDR (2013), NC3 Malaysia (2018), NC2 Lao PDR (2013), INC Myanmar (2012), NC2 Philippines (2014), NC4 Singapore (2018), NC3 Thailand (2018), NC3 Vietnam (2019), and the following chapters

The IPCC SROCC conveys that Southeast Asia is expected to experience more frequent, longer duration, wider coverage, and higher intensity marine heatwaves. Changes in wave heights arising from altered weather patterns as well as tidal amplitude changes due to sea level rise will influence coastal hazards differently in the region. Marine biomass production will be altered, affecting fish catch potential, with tropical oceans expected to experience the highest impacts threatening the livelihoods of economically vulnerable coastal communities. The redistribution of ocean resources is expected to increase the risk of conflicts and bring widespread challenges for fisheries governance.

1.3.2 Tackling Climate Change Risks

The latest National Communications (NC) submitted by AMS under the UNFCCC were reviewed to delineate their projected impacts and adaptation priorities (INC Myanmar 2012; NC2 Lao PDR 2013; NC2 the Philippines (2014); NC2 Cambodia 2015; NC2 Brunei Darussalam 2017; NC3 Indonesia 2017; NC3 Malaysia 2018; NC4 Singapore 2018; NC3 Thailand 2018; NC3 Vietnam 2019). These were then compared with the specific projected impacts that are relevant to Southeast Asia in the IPCC Special Reports to assess the readiness of the AMS in tackling the challenges. Five categories were delineated (Table 1.1). The highest level of readiness is where the region-specific projected impact mentioned in the IPCC Special Report is also mentioned in the NC, and adaptation measures are prioritized for the projected impact. A medium level of readiness is ascribed where the region-specific projected impact mentioned in the IPCC Special Report is not stated in the NC, but the associated sector is a priority for adaptation. A low level of readiness is given where the specific projected impact mentioned in the IPCC Special Report for the region is not stated in the NC nor is the associated sector a priority for adaptation. A low level of readiness is given when a specific projected impact is not a conventional phenomenon in several AMS and has to be investigated. A "not relevant" status is assigned where an impact is not applicable. For example, the projected impact for the marine sector is not applicable for Lao PDR, which is a land-locked AMS.

Global warming of 1.5 °C is expected to challenge Southeast Asia with heavy precipitation associated with cyclones. The AMS that are traditionally exposed to cyclones have a medium level of readiness to handle heavy precipitation associated with this hazard (Table 1.1). The NC of these AMS allude to the need for investment in disaster preparedness and early warning to handle impacts associated with heavy precipitation, particularly floods. Some AMS such as Brunei, Indonesia, Malaysia, and Singapore are not within the tropical cyclone belt but they have been occasionally subject to heavy rainstorms associated with squalls during a tropical cyclone. Thus, projected heavy precipitation associated with tropical cyclones merits further investigation in these countries. While adaptation to sea level rise is a priority for all

AMS (except land-locked Lao PDR), the level of readiness within these countries require improvement, specifically for communities exposed to land degradation from the combined impacts of both sea level rise and cyclones.

The impacts of 1.5 °C global warming include increased number of hot days and emergence of unprecedented climatic conditions, greater net reductions in yields and CO_2-dependent nutritional quality of rice, as well as the largest impact on economic growth compared with other regions. Generally, there is medium level of readiness for these projected specific impacts in Southeast Asia (Table 1.1). All AMS have identified projected temperature increase and its association with dry conditions and forest fires as well as potential impacts on human health, plants and livestock as an issue of concern. In addition, extreme events have also been given priority. However, none of the NC explicitly mention the specific potential impacts and associated adaptation measures for handling the increased number of hot days and potential emergence of unprecedented climatic conditions.

Several AMS such as Brunei Darussalam, Indonesia, Malaysia, the Philippines, Thailand, and Vietnam have prioritized adaptation measures to address yield reductions in rice due to increased temperature and salt water intrusion (Table 1.1). However, a similar level of readiness is not apparent in the NC of other AMS. The readiness to handle reduced yield in other types of food crops have not been adequately prioritized in many AMS. In addition, there is only medium readiness to handle the issue of CO_2-dependent nutritional quality of rice in the region. While all the AMS acknowledge the importance of climate change impact on economic growth in their respective NC, there is only a medium level of readiness to the projected impact as tropical Southeast Asia moves toward global warming of 1.5 °C. A recent review of Cambodia, Indonesia, Laos, Malaysia, Myanmar, the Philippines, Thailand, and Vietnam estimates that between 8 and 13% of the GDP in these countries could be at risk due to rising heat and humidity as well as exposure to extreme precipitation and flooding by 2050. The capital stock at risk in these countries is also expected to double, up to 1.5% with an estimated value of about USD 220 billion (MGI 2020).

All AMS except Cambodia have given high priority to the marine sector but there is only a medium level of readiness in adapting to specific impacts such as increased marine heatwaves, reduced biomass communities, ocean productivity, and potential for fish catch as well as threats to the livelihoods of vulnerable coastal communities (Table 1.1). The redistribution of ocean resources is expected to increase the risk of conflicts and bring widespread challenges for fisheries governance, but this is not explicitly mentioned in any of the NC of the AMS, indicating a need for further investigation of this issue in Southeast Asia.

1.3.3 Transitioning to a 1.5 °C World

Astute Adaptation: A recent review of National Adaptation Plans (NAPs) that have been formulated by multiple countries both developed and developing, reveal that the plans reflect a more global agenda with less emphasis on national and sub-national vulnerabilities and contexts (Alvesa et al. 2020). Despite different socioeconomic levels of development, diverse climate risks, and dissimilar vulnerability and readiness status among countries, the examined NAPs tend to focus on the same critical sectors and objectives. The review of climate change adaptation in AMS in the subsequent chapters of this book reflect a similar pattern. Furthermore, a disconnect is observed between the global agenda-driven national priorities and local causes of social vulnerability, which is site-specific and context-dependent. The experience in Indonesia, the Philippines, and Thailand indicate that a human capacity approach was better suited for understanding local level social vulnerability. Ecosystem-based adaptation and community-based adaptation that focus on ecosystem services and their socioeconomic benefits to people or community, through conservation, rehabilitation, enhancement of adaptive capacity, and skill development have been advocated to be more useful. While the focus on adaptation at the national policy level may obscure the complexity of adaptation at the sub-national scale, in many cases local level implementation, particularly in cities, offer the best potential for integrated adaptation and mitigation plans (Pereira and Hunt 2019; Alvesa et al. 2020). City planning in many AMS is generally short term, physically oriented, and unable to respond effectively to environmental change, with weak institutional capacity including inefficient legislation, regulations, and processes in the planning system (Chen et al. 2020). With increasing urbanization, proactive management of urban land expansion, especially in small and medium cities, will be critical for saving agricultural lands in peri-urban regions while creating equitable and affordable urban landscapes (Güneralp et al. 2020). An astute adaptation philosophy that takes a blended approach of integrating economic risks and prospects as well as natural resources and conservation, offers AMS an opportunity to create synergies between mitigation and adaptation, and implementing cost-effective climate actions. This is relevant for all sectors such as the built environment, infrastructure, economic expansion, natural resources, health, and human resource development.

Linking Climate and Disasters: The first step in adaptation to future climate change is to reduce vulnerability and exposure to present climate variability, and this involves disaster risk reduction. At the regional level, there have been proactive efforts in linking climate change adaptation and disaster risk reduction through a variety of mechanisms, specifically to improve collective capacity (CFE-DM 2019). There are several commitments to promote development in building a more resilient ASEAN Community with regard to disasters and climate change. However, within the AMS there are separate platforms for climate change and disaster risk management. Enhanced coherence and streamlining of efforts within AMS will serve to optimize administration and increase overall financial and human resource efficiency. The disaster risk reduction cycle offers a variety of entry points for integrating both climate change adaptation and mitigation, particularly at the local level (Fig. 1.3). On a conceptual level, climate change adaptation can be integrated within measures taken for disaster prevention, mitigation or preparedness. Entry points for climate change mitigation include the disaster recovery, prevention and mitigation phases,

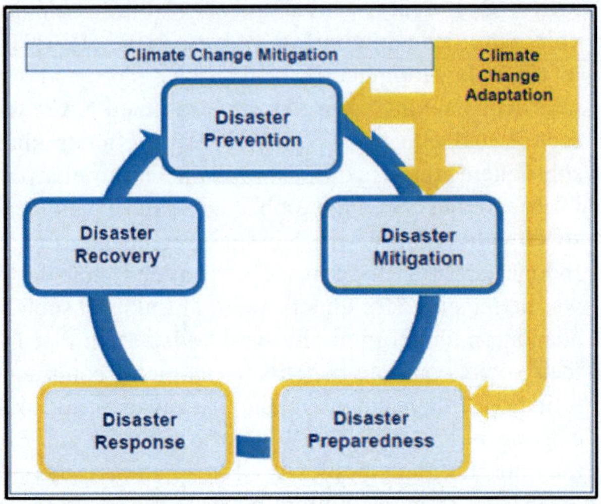

Fig. 1.3 The disaster risk reduction cycle offers entry points for both climate change adaptation and climate change mitigation. Climate change mitigation is feasible when building engineered structures

particularly when building engineered structures. A pilot is ongoing in Malaysia and Cambodia to develop local level disaster resilience plans, where disaster risk reduction measures including early warning are identified for climate driven hazards, while climate change adaptation actions are delineated to address long-term threats (http://ancst.org/idrc). The potential for shared benefits of adaptation is also being considered. The landscape for climate and disaster risks in the AMS was collectively costed at USD 86.5 billion in 2019, where 97% is associated climate driven hazards, primarily droughts, floods and tropical cyclones (ESCAP 2020). This cost is expected to increase with 1.5 °C global warming and the projected increase of climate extreme events in the region. The reality is that the AMS have to deal with costs associated with physical risks and liability risks due to climate driven hazards, in addition to expenditure associated with transition to a low carbon economy. The upscaling of local level synergies between climate change adaptation, climate change mitigation and disaster risk management offers AMS an opportunity for a more balanced approach to traverse this transition.

Amplified Engagement: Adaptation in the AMS is primarily driven by national governments and international institutions with lesser involvement from lower levels of government, non-government organizations, civil society organizations, academia and the private sector. There is also minimal collaboration on adaptation across AMS. The Ministry of Environment in most AMS play a central role in climate change policies and are well-positioned to mobilize adaptation planning. However, they appear to be constrained by existing institutional arrangements, such as conflicting mandates or fragmentation, or low political or public prioritization of climate change, which impede climate change adaptation at the local level. The primary climate change institution in several AMS such as Cambodia, Loa PDR, Thailand and Vietnam have strong linkages with their national disaster management organizations. This has facilitated the flow of information to the local level but many challenges remain, leaving the issue to be handled within the government domain. With the advent of extreme climate and increased risks of climate influenced hazards such as floods and landslides, among others, the AMS will be expected to strengthen gaps in disaster preparedness, building community resilience and mobilizing resources to handle the recovery, reconstruction, restoration and rehabilitation after disaster events. The experience of stakeholder engagement in several AMS have revealed that the merging of science-based data with community-level perspective has been most effective in filling knowledge gaps (Gustafson et al. 2017). This calls for an increased and stronger role for non-government players at the local level. Strong national coordination and amplified engagement of lower levels of government as well as deep and meaningful involvement of non-government organizations (NGOs), civil society organizations, academia and the private sector has been proposed as a way forward in Malaysia. Area- and context-specific stakeholder engagements have been advocated to systematically and continuously enhance adaptive capacity, build disaster resilience and integrate adaptation into policies, plans and strategies in a coherent manner into various sectors and at the relevant scale.

Inspired Innovation: A document analysis of two major databases of quality scholarly journals, the Web of Science (WoS) and Scopus, revealed that collectively, AMS generated a significant number of articles in Asia, between 2016 and mid-2020 (Fig. 1.4). Knowledge was generated on topics encompassing climate change adaptation, climate change mitigation and disaster risks (Fig. 1.5). Among the AMS, Indonesia was the highest knowledge generator, with over 2200 articles. Malaysia, Singapore, Thailand, Vietnam and the Philippines produced over 1000 articles each between 2016 and mid-2020. The generation of new knowledge plays an important role in tackling the negative impacts of climate change. The disclosure of research products in the form of quality scholarly products such as peer-reviewed indexed journals and adoption of open science strategies has a positive transformative effect on innovation (Jong and Slavova 2014). A recent assessment of multiple countries

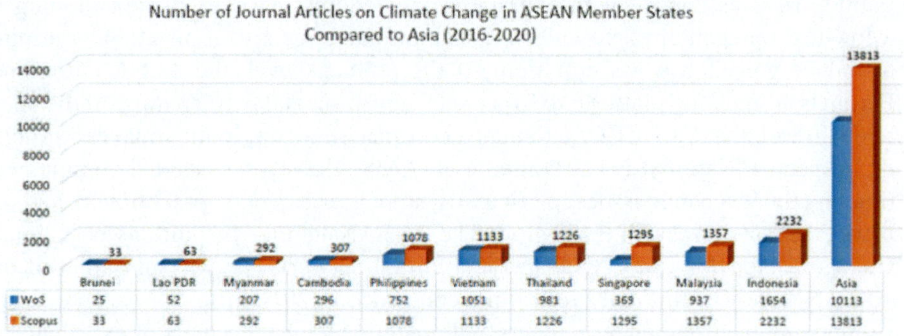

Fig. 1.4 The amount of journal articles generated on climate change in AMS compared with the total number of journals in Asia, between 2016 and mid-2020. *Source* Web of science, wos and scopus databases

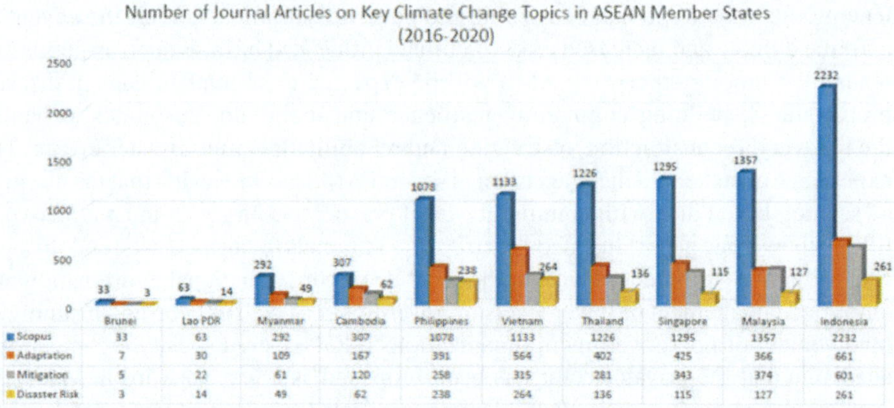

Fig. 1.5 The number of journal articles generated on climate change topics, primarily adaptation, mitigation and disaster risk in AMS is the highest in Indonesia with over 2200 articles, with Malaysia, Singapore, Thailand, Vietnam and the Philippines producing over 1000 articles each between 2016 and mid-2020. *Source* Scopus database

revealed the positive effect of innovation at the country level on the country's vulnerability to climate change (Dina et al. 2020). Irrespective of income levels, a high level of innovation outputs based on application of knowledge was found to have a positive effect in decreasing levels of a country's vulnerability to climate change. This finding indicates the importance for AMS to invest in increasing innovation, improving existing institutional arrangements to encourage innovation and promote open science strategies within their resource limits, to cope with climate change. Innovation offers an important new opportunity for inspiring AMS to tackle climate change adaptation and mitigation synergies, in addition to other instruments.

Smart Partnerships: Given the ominous prognosis for the region due to global warming of 1.5 °C, partnerships are critical for AMS in dealing with physical and liability risks associated with extreme events, addressing risks of transitioning to a low carbon economy and calling for ambitious targets to limit global warming to reduce greenhouse gas emission. At the political level, the ASEAN Regional Forum is an enabling platform to foster such governmental partnerships. At the operational level, the AWGCCC maintains coordination among policy makers through the respective National Focal Points in all AMS. However, sustained communication on climate change issues, particularly among researchers, practitioners and the private sector is weak in the region. The ASEAN Partner Institutions on Climate Change Adaptation (ASEANadapt) is a virtual network that was an output of the ASEAN-India Green Fund Project that concluded in 2016 (https://aseanadapt.org). Comprising like-minded researchers and practitioners from centers of excellence in AMS, the network promotes the exchange of information and joint actions to advance local level climate change adaptation. The potential of this network and other such similar groups have not been optimized by AMS. Similarly, the ASEAN Connectivity

Agenda is a high-profile initiative that covers physical, institutional and people-to-people linkages as articulated in the Master Plan on ASEAN Connectivity (MPAC) 2025 (Lim and Vehuliza 2020). The potential of the Connectivity Agenda as articulated in the MPAC has not been leveraged upon to develop the full scale of digital technologies that could foster, generate and maintain information that relate to climate change and disaster risk reduction to benefit the private sector and communities in the AMS. Moving forward, the establishment of an ASEAN Connectivity Hub on Climate Change could be explored under the direction of the AWGCCC, comprising ASEANadapt and all the centers of excellence from AMS as well as other groups with links to disaster management institutions, to advance climate change adaptation in the region. The Hub could serve as a digital nucleus for multi-disciplinary and multi-sectoral groups to enhance connectivity among researchers and practitioners in AMS, facilitate exchange of information, sharing of expertise, promote capacity and awareness building, and enhance coordination of stakeholders groups, particularly the private sector, at multiple levels in the region. This could support AMS to hasten the process of developing and implementing truly effective climate change adaptation, in this post-COVID-19 pandemic era. The Hub could also play a role in advancing means of monitoring and evaluating current adaptation measures in the region.

Greater Ambition: The AMS have cities and urban agglomeration that are densely populated with increasing dependence on private vehicles that is a major source of greenhouse gas emissions. Although car ownership per capita is still low, growing car ownership continues in urban areas (Chen et al. 2020). In this regard, the AMS have made a commitment to reduce the average fuel consumption per 100 km of new light-duty vehicles by 26% between 2015 and 2025, with strengthened fiscal policies to incentivize consumers to purchase efficient vehicles, striving towards a regional standard in the long term (ASEAN 2019). The energy intensity of the manufacturing, commercial and agriculture sectors that play a key role in the economic structures of Indonesia, Malaysia, the Philippines, Singapore, Thailand and Vietnam have been decreasing over the years (Setyawan 2020). This is in part a reflection of existing programs that are ongoing in the AMS (ACE 2017). The AMS are committed to further reduce energy intensity by 30% by 2025 and increase the component of renewable energy mix by 23% by 2025 (ASEAN 2019). It has been stressed that the commitment of AMS in climate change mitigation requires adequate capacity building, transfer of technology and financial support, and shall be aligned with national needs and priorities. An emerging priority area is the health sector. Recent studies using data from the AMS to explore the linkages of emissions, health and the economy, confirmed that air pollution from fossil fuel combustion has a hidden cost on public health (Farhad and Farzad 2020; Fong et al. 2020). Carbon dioxide and particulates (PM2.5) are major risk factors for lung cancer, which is a major cause of cancer-related death in AMS. Increasing use of renewable energy and higher health-care expenditure per capita tend to reduce the prevalence of lung cancer. This calls for greater ambition in the AMS to create a healthy environment, including transfer of fossil fuel subsidies to renewable energies, carbon tax for the power generation sector and increased investment in green energy projects. This post-COVID 19 pandemic era is an opportunity for accelerating transition to a low carbon economy (Box 1.1). Notwithstanding, all nations have to scale up their collective commitment to reduce greenhouse gas emission in an effort to limit global warming to 1.5 °C, and developed countries must meet climate finance obligations as pledged.

BOX 1.1

Post COVID-19 Reset: Building Resilience to Climate and Disaster Risks in Southeast Asia

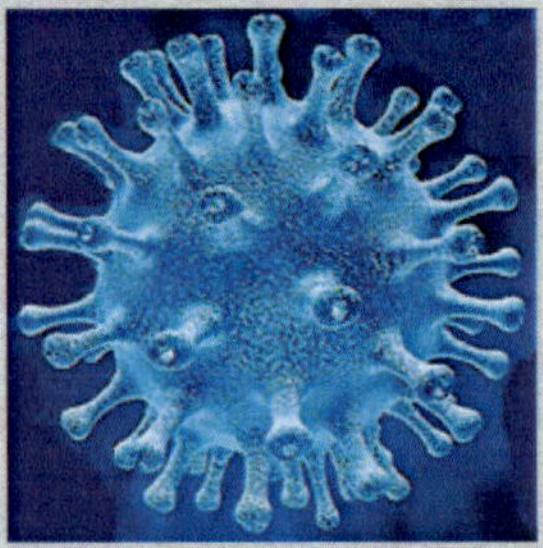

The COVID-19 pandemic has resulted in over half a million deaths globally and caused extensive social and economic impacts. In Southeast Asia, serious gaps have been exposed in social protection systems, delivery of public goods, weak health systems and income inequality, among others. Loss of jobs and livelihood, bankruptcy and other dire consequences have been reported in many sectors. Cascading effects include increased erosion of institutions, human rights violations and illegal fishing by foreign ships as government priorities focused on the pandemic itself and the challenges in humanitarian aid and disaster relief (Bennet et al. 2020; Ishiwatari et al. 2020; UN 2020). While the death toll in Southeast Asia is relatively lower compared with other regions, the recovery is expected to be slow due to the significant informal economy and large unbanked populations (Daly et al. 2020).

COVID-19 has brought to fore the importance of risk-informed decision making to manage future shocks, including increasing climate- and disaster-related shocks. The provision of scientific knowledge for understanding risks, strengthening risk governance, promotion of community engagement as well as strong coordination between the water, sanitation and health sectors have been found to be fundamental in flood disaster relief operations during the COVID-19 period (Ishiwatari et al. 2020). Companies that have high levels of environmental stewardship, social responsibility and governance track records are expected to perform better during the short, medium and long-term COVID-19 recovery phase (Johnstone-Louis 2020). The post-COVID-19 economic recovery era offers an extraordinary prospect for ASEAN Member States (AMS) to demonstrate greater ambition in tackling climate and disaster risks in the region. Priorities could be reset to use the stimulus packages to drive investments and behavioral changes that hasten the transition to a low carbon economy.

Policies that deliver on both economic and climate goals include investment in energy efficiency, renewable energy, education and training to address unemployment and structural shifts to de-carbonization, nature-based solutions and rural development as well as research and innovation (Hepburn et al. 2020). Ecosystem rehabilitation measures offer a high level of co-benefits by reducing carbon emission, enhancing carbon sequestration, preventing further erosion of biodiversity and reducing the risk of a future pandemic (Dinerstein et al. 2020). The involvement of local governments and non-state actors in COVID-19 recovery programs that target the community is effective in reducing economically vulnerable households (Gerard et al. 2020). The involvement of local governments and non-state actors is also critical to advance climate change adaptation and disaster risk reduction. There are many options and sectoral areas for investment which are of relevance to the AMS for implementing COVID-19 recovery programs (ADB 2020; Daly et al. 2020). Context- and area-specific measures with emphasis on co-benefits could serve to accelerate the transition of AMS to a low carbon economy while simultaneously building resilience to climate and disaster risks.

1.3.4 Strategic Initiatives on Climate Change

A critical element for climate management in AMS is to focus on coping with extreme events and disaster risk management using the best available scientific information at the local level, and in conjunction with all stakeholders. There is an opportunity for developing regional initiatives that are process-oriented flexible adaptation and taking a balanced approach to integrate climate change mitigation and other environmental concerns, via ecosystem-based adaptation and community-based adaptation, for example. In order to achieve a high level of impact on the ground, regional initiatives should be action oriented, have implicit capacity building, education and mainstreaming components.

Regional initiatives should also involve local policymakers right from the design stage as well as engage stakeholder participation from federal, state and local levels, in addition to universities, NGOs and civil society groups, among others. Homegrown private sector support could also be engaged where feasible but a majority of the AMS will require resources and technical support to implement the initiatives. The pilots should be multi-disciplinary, multi-sectoral and multi-consortium with the capacity to be replicated and upscaled. Initiatives should be sustainable and facilitate institutional change; once it is over, the mechanism and processes must go on after the project proponent steps out and the funding period is over.

A range of local level pilots have been proposed for all AMS within the chapters that follow, encompassing multiple sectors such as forest biodiversity, coastal, energy, urban, water resources, among others. Within each pilot, there is opportunity to develop local level tools and approaches that are both area specific and cross-cutting in nature. Examples include integration of disaster risk reduction and climate change adaptation, community-based adaptation, ecosystem-based adaptation, among others. Expanding on the findings of the ASEAN-India Green Fund Project and several regional strategic initiatives are recommended for advancing climate change adaptation in AMS.

ASEAN Pilot on Local Climate Change Adaptation: This constitutes the local level pilots highlighted in all the AMS (see Chaps. 2–11). The pilot has to be of a manageable size; consisting of a landscape, ecosystem, sub-watershed or a sub-basin, depending on the needs of the AMS. The pilot proposed should have national interest and make a difference on the ground i.e., at the community, city or ecosystem levels. This will ensure that the government has a "buy-in" into the project, and external funds serve as a leverage to enhance synergies. Challenges associated with the initiative relate to uncertainty associated with downscaled climate projections that require judicious use of derived findings for high investment decisions. Novel approaches are required for dealing with the anticipated impacts of climate change at the local level. Vulnerability to current climate has to be differentiated from vulnerability to long-term climate, and both approaches are equally important. Methods are available for both approaches. An inherent vulnerability index may be suitable for disaster risk reduction or short-term adaptation of development projects to current climate (i.e., water resources management, ecosystem-based adaptation approaches, etc.). For long-term resilience, the use of vulnerability index based on climate change-based scenario is more relevant.

Transboundary Initiatives on Climate Change: These encompass several cross-cutting issues that influence regional security. These include disrupted food supply chains, water resources in the Lower Mekong River Basin as well as ocean ecosystems that are projected to be emerging issues in the region. Forests in Southeast Asia are expected to be impacted by climate change. Hence, the aspect of carbon stock is of importance as carbon sinks could be impacted. Coping strategies and adaptation practices for addressing climate change impacts on forest, including production, should be taken into account in conducting simulation or modelling studies. Carbon credit mechanisms among the AMS, particularly on aspects of securing income or incentives out of carbon credits, is also an issue that requires capacity building and development of experts from the region.

Other Strategic Initiatives: These include the ASEAN Panel on Climate Change and ASEAN Weather Climate Advisory System. The purpose of the ASEAN Panel on Climate Change is to review the current state of science on a periodic basis in the region. An initial pilot such as a "mini IPCC" assessment could be supported, after which other mechanisms need to be explored to ensure its sustainability. The ASEAN Weather Climate Advisory System is intended to provide climate information to multiple stakeholders, particularly the agrarian communities such as farmers and fishermen of AMS. The focus is on near-term climate variability and it has to be led by the Meteorology Department of the respective AMS. There are already ongoing pilots in other regions that have been quite successful.

1.4 Concluding Remarks

Climate change and disaster risks is mainstreamed in the three major pillars of ASEAN, as reflected in the Political-Security Community Blueprint 2025, ASEAN Economic Community Blueprint 2025 and ASEAN Socio-Cultural Community Blueprint 2025, which complement the UN 2030 Agenda for Sustainable Development. As of 2019, all AMS have ratified the Paris Agreement and communicated their Nationally Determined Contributions (NDCs). The AMS have also periodically submitted their respective National Communications under the UNFCCC.

The projected impacts of global warming in Southeast Asia include higher levels of precipitation associated with tropical cyclones, greater net reductions in yields of maize, rice, and other cereal crops as well as CO_2-dependent nutritional quality of rice, increased number of hot days, and largest reduction of economic growth; all of which can be avoided by limiting global warming to 1.5 °C, rather than 2 °C or more. Overall, there is only a medium level of readiness in handling the projected impacts in the AMS; Brunei Darussalam, Indonesia, Malaysia, the Philippines, Thailand, and Vietnam have prioritized adaptation measures to address reduction in rice yields.

Astute adaptation involving a blended approach of integrating economic risks and prospects as well as natural resources and conservation at the local level including cities, promotes cost-effective climate actions. Linking Climate and Disasters by scaling up local level synergies between climate change adaptation, climate change mitigation and disaster risk management will help facilitate the AMS that have to deal with costs associated with physical risks and liability risks due to climate driven hazards, in addition to expenditure associated with transition to a low carbon economy. This is critical in a landscape where the cost of climate and disaster risks in the region is expected to increase with 1.5 °C global warming due to projected increase of climate extreme events.

Amplified engagement with stakeholders that take into account area and context specificities has been advocated to systematically and continuously enhance adaptive capacity, build disaster resilience, and integrate adaptation into policies, plans, and strategies. In addition to other instruments, investment in increasing innovation, improving existing institutional arrangements, and promoting open science strategies within respective resource limits of the AMS offers an important and new opportunity to cope with climate change. Smart partnerships could support AMS to hasten the process of developing truly effective climate change adaptation plans with multiple partners including the private sector, especially in this post-COVID-19 pandemic era. The establishment of a Climate Change Digital Hub could enhance connectivity among researchers and practitioners from multidisciplinary and multi-sectoral groups in the AMS. Greater ambition to transition away from fossil fuels is required in the AMS to create a healthy environment. Notwithstanding, all nations have to scale up their collective commitment to reduce greenhouse gas emission in order to limit global warming to 1.5 °C; developed countries must meet climate finance obligations as pledged.

Potential strategic initiatives that could benefit AMS include those that develop local level tools and approaches that are both area-specific and cross-cutting in nature, encompassing integration of disaster and climate risks, community-based adaptation, and ecosystem-based adaptation, among others. Initiatives should be action oriented and have implicit capacity building, education and mainstreaming components, involve local policymakers right from the design stage as well as other specific stakeholders to be impactful on the ground. While homegrown private sector support is an option, many of the AMS will require resources and technical support to implement the initiatives.

Acknowledgements The ASEAN-India Green Fund is acknowledged for supporting the project on "Enhancing Local Level Climate Change Adaptation in Southeast Asia" (Phase 1), which provided part of the findings documented within the chapters of this book. Ms. Natalia Derodofa of the ASEAN Secretariat and all National Focal Points of the ASEAN Working Group on Climate Change (AWGCC) as well as Mr. Rocky Pairunan and Dr. Raman Letchumanan are thanked for facilitating this project.

References

ACE (2017) The 5th ASEAN energy outlook ASEAN centre for energy. Jakarta, Indonesia

ADB (2020) COVID-19 recovery: a pathway to a low-carbon and resilient future. Asian Development Bank, Manila, p 8. https://www.adb.org/publications/

ADRMI (2010) ASEAN disaster risk management initiative: synthesis report on ten ASEAN countries disaster risks assessment. ASEAN Secretariat, Jakarta, p 149

Alvesa F, Filho WL, Casaleirob P, Nagye GJ, Diazf H, Al-Amin AQ, de Andrade GJBSO, Hurlberti M, Farooqj H, Klavinsk M, Saroarl M, Lorencovam EK, Sureshn J, Soareso A, Morgadoo F, O'Harep P, Wolfq F, Azeiteiroo UM (2020) Climate change policies and agendas: facing implementation challenges and guiding responses. Environ Sci Policy 104(2020):190–198

ASEAN (2019) ASEAN joint statement to the United Nations climate action summit 2019. New York, 23 Sep 2019, p 2. https://www.asean.org

ASEAN (2020) The ASEAN Charter. ASEAN Secretariat, Jakarta, p 60. https://www.asean.org

Bennett NJ, Finkbeiner EM, Ban NC, Belhabib D, Jupiter SD, Kittinger JN, Mangubhai S, Scholtens, J, Gill, D, Christie P (2020) The COVID-19 pandemic, small-scale fisheries and coastal fishing communities. Coast Manag. https://doi.org/10.1080/08920753.2020.1766937

CFE-DM (2019) ASEAN disaster management reference handbook. Center for excellence in disaster management & humanitarian assistance. Honolulu, p 63

Chen TL, Chiu HW, Lin YF (2020) How do east and southeast Asian cities differ from western cities? a systematic review of the urban form characteristics. Sustainability 12(2423):1–19. https://doi.org/10.3390/su12062423

Daly E, Dass K, Yeoh R (2020) Reimagining emerging ASEAN in the wake of COVID-19. McKinsey & Co. 2:5

Dina MA, Aleksey M, Angie MA (2020) Vulnerability to climate change: are innovative countries in a better position?. Res Int Bus Financ 51:101098. https://doi.org/10.1016/j.ribaf.2019.101098

Dinerstein E, Joshi AR, Vynne C, Lee ATL, Pharand-Deschênes F, Franca M, Fernando S, Birch T, Burkart K, Asner GP, Olson D (2020) "Global safety net" to reverse biodiversity loss and stabilize Earth's climate. Sci Adv 6:1–13. http://advances.sciencemag.org

ESCAP (2020) The disaster risk scape across Asia-Pacific: pathways for resilience, inclusion and empowerment. Asia-Pacific disaster report 2019. United Nations economic and social commission for Asia and the Pacific (ESCAP). Bangkok. https://www.unescap.org/publications/asia-pacific-disaster-report-2019

Farhad TH, Farzad TH (2020) The impacts of air pollution on health and economy in Southeast Asia. Energies 13(1812):1–15 https://doi.org/10.3390/en13071812

Fong LS, Salvo A, Taylor D (2020) Evidence of the environmental Kuznets curve for atmospheric pollutant emissions in Southeast Asia and implications for sustainable development: a spatial econometric approach. Sustain Dev 1–16. https://doi.org/10.1002/sd.2097

Gerard F, Imbert C, Orkin K (2020) Social protection response to the COVID-19 crisis: options for developing countries. https://academic.oup.com/oxrep/advance-article/doi/.org/10.1093/oxrep/graa026/5899012

Güneralp B, Reba M, Hales BU, Wentz EA, Seto KC (2020) Trends in urban land expansion, density, and land transitions from 1970 to 2010: a global synthesis. Environ Res Lett 15(2020):1–13. https://doi.org/10.1088/1748-9326/ab6669

Gustafson S, Cadena AJ, Ngo CC, Kawash A, Saenghkaew L, Hartman P (2017) Merging science into community adaptation planning processes: a cross-site comparison of four distinct areas of the Lower Mekong Basin. Clim Change 149(3):91–106. https://doi.org/10.1007/s10584-016-1887-7

Hepburn C, O'Callaghan B, Sterni N, Stiglitzii J, Zenghelisiv D (2020) Will COVID-19 fiscal recovery packages accelerate or retard progress on climate change? https://academic.oup.com/oxrep/advance-articlehttps://doi.org/10.1093/oxrep/graa015/5832003

INC Myanmar (2012) Myanmar's initial national communication under the United Nations framework convention on climate change. Ministry of environmental conservation and forestry. Nay Pyi Taw, p 268

Ishiwatari M, Koike T, Hiroki K, Toda T, Katsube T (2020) Managing disasters amid COVID-19 pandemic: approaches of response to flood disasters. Prog Disaster Sci 6. https://doi.org/10.1016/j.pdisas.2020.100096

Johnstone-Louis M, Kustin B, Mayer C, Stroehle J, Wang B (2020) Business in times of crisis. Oxf Rev Econ Policy. https://doi.org/10.1093/oxrep/graa021/5898860

Jong S, Slavova K (2014) When publications lead to products: the open science conundrum in new product development. Res Policy 43(4):645–654

Lim CC, Vehuliza TI (2020) Shaping ASEAN's human capital development with connectivity. The ASEAN 2:26–27

MGI (2020) Climate risk and response in Asia-Future of Asia: research preview. McKinsey Global Institute. https://www.mckinsey.com/mgi

NC2 Brunei Darussalam (2017) Brunei Darussalam's second national communication under the United Nations framework convention on climate change. Energy and industry department, The Prime Minister's Office, Bandar Seri Begawan, p 49

NC2 Cambodia (2015) Cambodia's second national communication under the United Nations framework convention on climate change. General secretariat, national council for sustainable development/ministry of environment. Kingdom of Cambodia, Phnom Penh, p 94

NC2 Lao PDR (2013) Lao PDR's second national communication under the United Nations framework convention on climate change. Government of Lao PDR, Vientiane, p 124

NC2, Philippines (2014) The Philippines's second national communication and second biennial update report under the United Nations framework convention on climate change. Government of The Philippines, Manila, p 102

NC3 Indonesia (2017) Indonesia's third national communication under the United Nations framework convention on climate change. Ministry of environment and forestry. Jakarta, p 219

NC3 Malaysia (2018) Malaysia's third national communication and second biennial update report under the United Nations framework convention on climate change. Ministry of energy, science, technology, environment and climate change. Putrajaya, p 337

NC3, Thailand (2018) Thailand's third national communication under the United Nations framework convention on climate change. Ministry of natural resources and environment, Bangkok, p 162

NC3 Vietnam (2019) Third national communication of Vietnam under the United Nations framework convention on climate change. Ministry of natural resources and environment. Hanoi, p 124

NC4 Singapore (2018) Singapore's fourth national communication and third biennial update report under the United Nations framework convention on climate change. National Environment Agency, Singapore, p 142

OECD (2019) Economic outlook for Southeast Asia, China and India 2019: towards smart urban transportation. The organisation for economic co-operation and development. Paris

Pereira JJ, Hunt JCR (2019) Climate science & technology in Asia: current status and future needs. Report of the workshop on status of climate science and technology in Asia, 15–16 (November 2018) Kuala Lumpur. LESTARI Publisher, Bangi, Malaysia, Malaysia

Setyawan D (2020) Economy-wide energy efficiency using a comprehensive decomposition method. Global J Environ Sci Manage 6(3):385–402

Shi X (2016) The future of ASEAN energy mix: a SWOT analysis. Renew Sustain Energy Rev 53:672–680

The ASEAN Post (2019) ASEAN growth slower than forecasted. The ASEAN post 30 Dec 2019. https://www.theaseanpost.com. Accessed 1 July 2020

UN (2020) Policy brief: the impact of COVID-19 on South-East Asia. United Nations, New York, p 29

UNDP (2019) Human development report 2019. United Nations development program, New York, p 366

WPR (2020) World population review 2020. https://worldpopulationreview.com/country-rankings/asean-countries. Accessed 1 July 2020

Chapter 2
Climate Change Adaptation in Brunei Darussalam

Rahayu Sukmaria Sukri and Dennis Wah

Abstract Brunei Darussalam's greatest vulnerability to climate change probably lies in the energy sector, which sees the problem of having to move away from the use of, and reliance on, fossil fuels and petroleum products, the country's main source of income. This is followed by vulnerabilities in the water and agriculture sectors. Although adaptation actions for climate change are mainly focused on responding to the impacts of climate change, it can also include preventative measures to slow down the progression of climate change and mitigation measures to reduce the effects. There is need for more regionally relevant studies to be carried out on the potential impacts of climate change in the Bornean region. This would be invaluable for accurately predicting how climate change will affect the region and provide better information in preparation for climate change. A local level pilot focused on agriculture is proposed, to be located in any of the 66 suitable Agricultural Development Areas of Brunei.

Keywords Climate change adaptation · Climate mitigation · Vulnerability · Disaster risk · Brunei Darussalam

2.1 Introduction

While relatively safe from most natural disasters, Brunei Darussalam is not exempted from the effects of climate change. As global surface mean temperature and weather variability increase, Brunei is vulnerable to thunderstorms, monsoon floods, landslides, loss of forestry and biodiversity, loss of food security, resurgence of diseases for public health, as well as haze and forest fires during the dry season (ASEAN Inter-Parliamentary Assembly 2012; Brunei Darussalam's Second National Communication 2017). In relatively recent years, Brunei has also

R. S. Sukri (✉) · D. Wah
Universiti Brunei Darussalam, Bandar Seri Begawan, Brunei
e-mail: rahayu.sukri@ubd.edu.bn

D. Wah
e-mail: dennis.ting@ubd.edu.bn

J. J. Pereira et al. (eds.), *Climate Change Adaptation in Southeast Asia*, Disaster Risk Reduction, https://doi.org/10.1007/978-981-16-6088-7_2

experienced increased heavy rains which have caused flooding and landslides (Baines 2011; Loo et al. 2014) and also seasonal haze resulting from forest fires in neighboring countries (Shams et al. 2019). While these disasters are not necessarily brought about by the impact of climate change (The Guardian 2014; Velde 2015), the likelihood of their occurrence and persistence is exacerbated by it (Bengtsson and Hodges 2006; Manton et al. 2001; Page et al. 2009). Similarly, El Niño Southern Oscillation (ENSO) events have also been reported in Brunei (Brunei Darussalam Meteorological Department 2015; Othman 2016), which can further contribute to these disasters. The total annual average rainfall for Brunei is 2,300 mm with the temperature ranging from 23 °C to 32 °C. The hottest months are from February to April, whereas the coolest months are from late October to early January, and from May to July due to the Northeast Monsoon and Southwest Monsoon. Downscaled modelling predicts that Brunei is expected to experience more warming and less frequent but more intensified high rainfalls in the future (Hasan et al. 2019).

The country has a small surface area with a small population compared to other Association of Southeast Asian Nations (ASEAN) member states, and a population density of 79.3 persons/km^2 spread across four districts. It is estimated that about 77% of the country is now urbanized (Ndah and Odihi 2017). Likewise, most (if not all) of Brunei is situated on the coast of Borneo and thus, a majority of the Bruneian population live on coastal areas which are susceptible to rising sea levels (BruDirect 2004; Michener et al. 1997; Nicholls et al. 1999; Silvestre et al. 1992). Brunei is vulnerable to hazards such as earthquakes, cyclonic storms, floods, landslides, seasonal forest fires, and smoke/haze (Ndah and Odihi 2017; INFORM 2017). The haze problem will also be more prevalent with altered forest fire frequency, intensity, and timing (Dale et al. 2001). To date, floods remain the main disaster that have affected the country dearly, followed by landslides; even the occurrence and impacts of landslides have remained largely under-reported in international disaster databases (Ndah and Odihi 2017).

Brunei Darussalam does not yet have a dedicated plan for dealing with climate change, and this is reflected in its current policies towards, climate change adaptation and mitigation. Much of the local climate change efforts of Brunei Darussalam are of a subsidiary nature derived from other policies focused on issues such as sustainable resource regulations or forest protection (Forestry Department 2010; and other examples: Water Supply Act, Land Acquisition Act, Land Code, Town and Country Planning Act, Forestry Act, Fisheries Order 2009, Wild Fauna and Flora Order 2007, Wildlife Protection Act, Ninth National Development Plan 2007–2012, Heart of Borneo Initiative). This report aims to produce a clear overview of Brunei's national status on adaptation, vulnerable sectors and regions, and its needs and priorities against climate change. Taking this information into account, several local level pilots that are of most relevance to Brunei are proposed.

2.2 National Perspective

2.2.1 National Arrangements and Institutions

While not dedicated solely to climate change, the most relevant national documents of Brunei Darussalam on adaptation actions to climate change are summarized in this section. Since adaptation actions for climate change are mainly focused on responding to the impacts of climate change, it can also include preventative measures to slow down the progression of climate change and mitigation measures to reduce the effects.

The Disaster Management Order (2006) started up the National Disaster Council (NDC) and established the National Disaster Management Centre (NDMC) in Brunei Darussalam (Abidin and Ahmad 2006). The Disaster Management Order (DMO) is concerned with the course of possible actions for managing all disasters and their potential adverse effects, i.e., plans for mitigating, preventing, preparing for, responding to and recovering from disasters. Under this order, disasters are defined as: "A serious disruption in a community, caused by the impact of an event, that requires a significant coordinated response by Brunei Darussalam to help that community recover from that disruption." A disruption in this case includes but is not limited to: (i) loss of human life, or illness or injury to humans; (ii) widespread or severe loss or damage to property; or (iii) widespread or severe damage to the environment. "A disruption may occur naturally or be caused by human acts or omissions" (Abidin and Ahmad 2006). Climate change covers all three listed definitions of a disruption, and hence is directly applicable as a disruption under the DMO. Indeed, it can be seen later in the implementation section of the report that many of the adaptation actions under the NDMC, and hence the DMO, are applicable as climate change adaptation actions.

The Heart of Borneo (HoB) Initiative is a trilateral initiative between the governments of Brunei Darussalam, Indonesia (Kalimantan) and Malaysia (Sabah and Sarawak) that transcends national boundaries (Heart of Borneo 2012). Its objective is to minimize forest degradation and the associated loss of biodiversity and ecosystem services while also enabling effective and informed conservation, and sustainable development of Borneo's tropical rainforests, and hence improving the welfare of the island's inhabitants. The HoB area covers approximately 30% of the island of Borneo's land area: more than 22 million hectares of tropical rainforest across the three partnering countries (Heart of Borneo 2012). While not intended as a climate change adaptation plan, the HoB Initiative is relevant nonetheless, since protecting Borneo's rainforests is a good way to mitigate the effects of climate change, which in turn will determine the effectiveness of Brunei's adaptation actions. Under the Brunei HoB Project Implementation Framework (PIF) (Department of Agriculture 2013), four strategies and 36 interventions have been formulated, some of which are of high relevance as mitigation and adaptation measures for climate change. Under Strategy 1, the preservation of the Belait peat swamps is highlighted as an important mitigation strategy for climate change as peat swamp forests are recognized as major carbon sinks. Brunei's peat swamp forests are currently the most intact in Borneo, and therefore of global importance. The Brunei HoB PIF also recommended that studies on the potential of carbon offsets, as well as Brunei's forests as a model of carbon sequestration in tropical forests, be implemented. The implementation of such efforts by Brunei Darussalam is expected to bring global recognition and provides an avenue for Brunei Darussalam to undertake its responsibility under the United Nations Framework Convention on Climate Change (UNFCCC).

The Brunei Darussalam National Council on Climate Change is co-chaired by the Minister of Development and the Minister of Energy and Industry at the Prime Minister's Office (Brunei Times 2011). The Council aims to formulate and implement national climate change mitigation and adaptation policies and strategies. The Brunei Darussalam Strategic National Action Plan (SNAP) for Disaster Risk Reduction 2012–2025, is a national document that illustrates Brunei's global commitment in ensuring a safer, disaster-resilient country and community (Brunei Darussalam's Second National Communication 2017). It was designed to cover the five priority actions of the Hyogo Framework for Action (HFA) (UNISDR 2007). The SNAP is a cross-sectoral action plan that enables both the consolidation of disaster risk reduction (DRR) activities and the reaffirmation that DRR is the responsibility of all citizens and sectors, and hence the reaffirmation that everyone is a stakeholder. The national and cross-sectoral nature of this action plan allows all significant ongoing initiatives to be mapped out with relative ease and subsequently any gaps in information and technology to be easily identified. Disaster risk reduction shares a common goal with climate change mitigation and adaptation: to reduce vulnerability of communities to weather and climate hazards, and to achieve sustainable development. Hence, climate change mitigation measures are integral to the SNAP. Under the SNAP, public awareness on the negative impacts of climate change is being promoted through education, workshops and public open days. The building codes in the country are also being supplemented with disaster and climate-risk mitigation measures. Existing policies and guidelines relating to DRR and prerequisites for drafting new laws and regulations are also constantly being reviewed.

2.2.2 Ongoing Projects

There are several projects that are being executed in Brunei which are not exclusively focused, but are relevant, to climate change adaptation and mitigation. These include the following:

- Tenaga Suria Brunei is a 1.2 MW large-scale photovoltaic power generation demonstration in Seria. This project, which is co-sponsored by the Mitsubishi Corporation, aims to evaluate the viability of utilizing solar power in the country (Cheong 2010).

- The Energy Efficiency and Conservation Initiative aims to improve the overall energy efficiency of Brunei via the switch to more efficient appliances (Le Quéré et al. 2009; Masli 2011). Under this initiative, there are also plans to provide detailed energy ratings on appliances to help promote buyer awareness on energy conservation; extend the concept to fuel economy regulation via the use of the mentioned energy ratings on cars and other larger devices; and introduce tax reductions on products/technology which are energy efficient, and also bans on energy-inefficient products.
- A rainfall pattern prediction system has been set up in the Brunei Darussalam Meteorological Department, Ministry of Communications, which is able to send warning messages to the relevant authorities when it predicts periods of extreme precipitation (Brunei Darussalam Meteorological Department 2015). This warning message is eventually relayed to the occupants residing in the relevant areas.
- The Department of Water Services has a Water Conservation Unit dedicated to promoting water-use awareness among the local populace. The objective of this unit is to help the people adopt a water conservation mindset and to try to mitigate the already high per capita water usage of the country. The unit is also responsible for holding seminars within the department to further boost the staffs' capacity for water conservation.
- All flood adaption and mitigation projects in Brunei can be considered as on-going pilots since they are constantly undergoing updates. Each project starts in a small test area and are then eventually adopted by all flood-prone areas once proven effective.
- The Green Building Initiative (Public Works Department 2015) aims to encourage a green approach to energy use behavior, e.g., limiting the number of operational elevators, temperature, and operating time regulations for air conditioning, etc.
- Land matchmaking is a practice that is encouraged among local farmers. It is where the type of land that is developed for agriculture determines the type of crops cultivated. For example, for peat lands, local variety of paddy are grown since the area will be naturally irrigated. Food crops that form sturdy canopies are recommended to be grown on hilly areas where the soil is especially prone to erosion (some degree of protection is required during the early stage of growth). Developed lowland areas are used to grow vegetable varieties that do not form strong canopies. Cultivated plots are also encouraged to have limited fertilizer and insecticide use.
- The use of local paddy varieties (Adan, Pusu, and Bario) is encouraged among local farmers. These local paddy varieties are selected based on their low requirement for water and their resilience to climate change.
- Paddy fields that are rainfed (fields very dependent on rain for irrigation) are encouraged to utilize an alternate drying and waiting watering scheme. This regimented water use increases the efficiency of paddy cultivation per unit water. However, this practice is done more due to water shortages rather than as a measure for climate change adaptation.

- Public health surveillance is continuously carried out to assess risks to the public and ensure the well-being of the population. In response to disease outbreaks and other potential risks to public health, the Ministry of Health often issues public notifications, alerts or health advisories to promote public awareness and empower the public in protecting their health and well-being.
- Carbon stock enhancement has been carried out since as early as 1992. Land rehabilitation, enrichment, and forest plantation are the main forms of carbon stock enhancement in Brunei. Land rehabilitation involves the planting of new trees in wasteland and degraded areas in order to improve its carbon storing potential. Land enrichment on the other hand involves the planting of timber species in protected areas in the hope of maximizing the biodiversity of the land. Forest plantations are patches of forests of relatively low value that are used for a continuous supply of timber. This is done by consistently planting and harvesting timber species on the land. Forest plantations are aimed at minimizing deforestation pressures on the other types of forest reserves, which tend to have relatively higher biodiversity values.
- A database of herbal vegetation and coral reef species (through the Marine Protected Areas) in Brunei is currently being compiled. Similarly, a compilation of orchid species is also in the works. The aim of these projects is to catalogue the wealth of plant species present in Brunei in hopes of attracting botanists and nature enthusiasts and hence promote green tourism. This is a pilot for establishing a non-destructive commercial infrastructure via rainforest conservation.

2.2.3 Key Measures and Challenges

Community Empowerment: The NDMC has been actively engaging local communities and conducting outreach programs and activities since its inception (Abidin and Ahmad 2006). The Community-Based Disaster Risk Management (CBDRM) was launched in 2010 (ASEAN Inter-Parliamentary Assembly 2012) and is aimed at empowering the communities of disaster prone areas, helping them to achieve higher levels of preparedness for disasters. Communities are taught to identify (potential) hazards, their associated risks, and ways to mitigate them. Additionally, interactive capacity-building lessons were also held in the form of basic life-saving drills which included fire-fighting and first aid. Since its launch, the program has been carried out in seven villages throughout Brunei. Similarly, the Brunei HoB Initiative has also been organizing conferences where academics and other stakeholders are brought up to speed on the current efforts of the Initiative, and also to voice their opinions or concerns about the HoB Initiative (HoB Trilateral Conference 2015, pers. comm., 13 April). These conferences provide a platform for stakeholders to aid in the effort to conserve Brunei's rainforests and to promote sustainable development within Brunei's HoB area.

Disaster Mitigation Structures and Improved Infrastructure: Between 2007 and 2012, multiple coastal protection projects were started by the Department of Drainage and Sewerage, Ministry of Development, which along with the continued plans on improving flood mitigation were part of Brunei's then-current National Development Plan (Sadikin 2014). These projects include, but are not limited to, the building of retaining walls and tidal valves to allow excess storm water to drain freely (through automatic regulation of water levels by pumping stations) and the increased elevation and expansion of roads in flood-prone areas. More recently, work is being done on both the maintenance of the country's main drainage system (storm and connectivity drains) and the construction of new structures such as detention ponds and concrete drains that would further improve overall drainage (Wong 2015). These mitigating efforts were initiated in May 2014 and are expected to be completed in October 2018 (Airudin 2014), with low-lying and flood-prone areas being prioritized. The appropriate surrounding rivers of these areas will also be deepened and widened via dredging as an extra precaution. Currently, detention ponds in Brunei cover a total area of 7.0 ha and can be as deep as 4.0 m (Airudin 2014).

Improved Disaster Data Collection: Prior to the establishment of the NDMC, national responses to disasters were executed on an ad hoc basis and hence adaptation measures were carried out by individual sectors based on the nature of the disaster. Therefore, while the following adaptation measures may not have originated from the NDMC, they are, nevertheless, part of the adaptation actions which Brunei is implementing (ASEAN Inter-Parliamentary Assembly 2012). The Brunei Darussalam Meteorological Department, Ministry of Communications are currently upgrading their climate database system that will allow for the collection of better quality data, which are in accordance with the World Meteorological Organization's standards. The new system will also increase the efficiency and ability at which researchers can collect and archive weather and climate forecast data. Universiti Brunei Darussalam (UBD) is currently utilizing climate change models in the form of high-resolution weather prediction models to help increase paddy/agricultural efficiency (JASTRE 2014, pers. comm., 15 April). Studies like these can benefit greatly from higher quality data, producing relatively more accurate and reliable results. A National Hydrology Centre has been set up by the Department of Drainage and Sewerage, Ministry of Development in order to maximize the value of the country's hydrometric data. The center will be providing better records on hydrometric factors to help improve government response to future floods via forecasting and real-time flood information. While not necessarily established as an adaptation plan for climate change, the Hydrology Centre's data records are an invaluable asset for climate change researchers in Brunei. These records can not only provide researchers with historical rain and flood patterns which can be used to model the effects of climate change in the region but also a location map of the areas that are most at risk. This is exactly what UBD is currently doing with its flood simulation model "FloodSim" (JASTRE 2014, pers. comm., 15 April). The "FloodSim" is used to help predict the frequency and intensity of extreme events such as changing sea levels, monsoon variability, forest fires, etc., and hence used to inform decision makers on appropriate disaster management initiatives.

Energy Efficiency: To mitigate the effects of climate change, the Brunei Government have launched various measures and initiatives for energy efficiency. The Ministry of Development's Green Building Initiatives (GBI) was launched in March 2012 (Anon 2015a). The GBI aims to highlight the significance of Green Building by applying "greener" approaches to the Ministry of Development Building and several buildings of the departments under the Ministry, such as the Public Works Department. The objectives of the initiatives are to save energy, promote safety, and reduce carbon emission that eventually helps the local economy in terms of exporting more instead of domestic consumption. Under the GBI, simple steps toward energy efficiency are taken, such as increasing the temperature of the air-cooling system in the buildings, setting the air conditioners to 23 °C; and at Public Works Department itself, the Department has adopted a "No Lift Day" for every Saturdays to save energy; and reminding the staff to switch off their electrical appliances before leaving the office and when not in use. Additionally, the Green Building Council, Brunei Darussalam (GBC Brunei) is currently working with the Ministry of Development on formulating a green building rating tool (Azaraimy 2015). The Ministry of Development, in collaboration with the Energy Department at the Prime Minister's Office, have developed the Energy Efficiency and Conservation (EEC) Guidelines for non-residential buildings. These guidelines take into account the methods and concepts in terms of green design, construction, operation, and maintenance of new and existing buildings as well as energy saving and water conservation. The Guidelines also outline various aspects such as best practices and proper installation techniques in addition to cost-effective and sustainable usage.

Research and Education on Climate Change: As the oldest and biggest higher education institution in Brunei Darussalam, UBD is responsible for the bulk of research on climate change. In addition to research on climate change modeling, UBD is also investigating the effects of climate change on Brunei Darussalam's biodiversity and ecosystems. The Biodiversity and Environmental Research of UBD, through UBD's International Consortium of Universities for the Study of Biodiversity and the Environment Initiative (iCUBE), organized the iCUBE Gastropod Thermal Biology and Climate Change Workshop on 8–11 December 2014 (iCUBE 2014). The objectives of the workshop were to draw attention to the threat of future climate change on natural tropical ecosystems; disseminate knowledge, promote awareness, and understanding of regional environmental and climate change issues, and their impacts on tropical ecosystems; strengthen existing and create new international links between Brunei Darussalam, and other countries through people-to-people interactions, sharing of information and knowledge for better mutual understanding of environmental and climate change issues; establish a platform for the discussion of climate warming vulnerability of tropical animals; as well as to work on specific conceptual outcomes that incorporate broader taxonomic and ecological inclusion into existing models for climate warming vulnerability, with special reference to gastropods. Additionally, UBD has set up a 25 ha permanent forest plot at the Kuala Belalong Field Studies Centre in collaboration with the Centre for Tropical Forest Science (CTFS) to conduct research that will provide long-term scientific data on environmental change, and to allow for long-term monitoring and assessment on the impacts and vulnerabilities of climate change on forest biodiversity and health of forest ecosystems. Researchers at UBD also conduct various research projects on climate change issues, which are translated into part of UBD's teaching curriculum.

Heart of Borneo Corridor Initiative: A trilateral, trans-boundary protection and conservation corridor is currently in the works under the HoB Initiative. The Heart of Borneo (HoB) Corridor Initiative previously referred to as the "HoB Spine," will serve as the ecological link connecting the conservation areas of Brunei Darussalam, Malaysia (Sarawak and Sabah), and Indonesia (Kalimantan). The HoB Corridor Initiative, which was proposed and subsequently spearheaded by Brunei Darussalam, will span across Sungei Ingei and Ulu Temburong National Park in Brunei Darussalam to Gunung Mulu National Park, the Kubuaan Puak forest management zone, and Pulung Tau National Park in Sarawak and into Indonesia's Kayan Mentarang National Park (Anon 2015b; Heart of Borneo 2012). It is hoped that the establishment of this trans-boundary corridor will lead to improved cooperation between both the people and governments of the partnering countries via improved information sharing and joint-effort opportunities (Anon 2015b; Heart of Borneo 2012). Based on a projected temperature increase of 2 °C by 2050 (Johnson 2012), Borneo will be severely affected by climate change through increased flooding, forest fires, and rising sea levels (Johnson 2012). Thus, in the context of climate change, the HoB Corridor Initiative is expected to contribute significantly to climate change adaptation. Climate change may cause species loss as habitats become unsuitable, resulting in shifts in species ranges. For example, up to 69% of current Orang Utan habitats are expected to become unsuitable by 2080 under current scenarios of climate change (Wich et al. 2015). The ecological links between the HoB countries will therefore potentially allow species to migrate to other potentially suitable areas within the Corridor, thus lowering the risk of complete species extinction from Borneo. Additionally, the trans-boundary corridor will play an important role in ensuring secure and healthy watersheds (Chan 2015), as well as ecosystem integrity in the face of climate change (Johnson 2012).

Challenges: The effectiveness of Brunei's climate change adaptation measures has been reported to be limited by the difficulty of changing the people's mindset, and the monitoring and evaluation of current adaptation measures (JASTRE 2014, pers. comm., 15 April). To counter the mindset problem, Brunei is currently trying to phrase climate change and its negative impacts in a more personal way to its people in hopes of getting them to adopt a relatively greener mindset. Additionally, a network of focal points appointed from relevant Government agencies has been suggested as one of the adaptation actions for climate change in Brunei Darussalam. There are other limiting factors such as limited expertise and resources, lack of dedicated national climate change adaptation reports or publications, and lack of regulatory framework on climate change adaptation policies. Steps are being taken to remedy these issues. Monitoring and evaluating the progress of climate change adaptation measures is a very subjective topic (e.g., how effective are the anti-disaster measures? How many adaptation measures are required? How prepared are the people?), and hence it is very difficult to judge how effective any of the adaptation measures actually are. With improved expertise and technology, it is hoped that simulation modelling, such as "FloodSim" can be used in the near future to help evaluate the effectiveness of Brunei's adaptation measures.

2.3 Adaptation Priorities

2.3.1 Key Sectors

Direct consultation with the relevant sectors revealed that all stakeholders share three main challenges for effective adaptation: improved collaboration, workforce, and data collection systems. All sectors agreed that current inter-sectoral collaboration and teamwork efforts are lacking and can be improved through improved communications. This lack of communication is also encountered when dealing with data collection and recording, where the different sectors believed that clearer and more structured guidelines would benefit them greatly, i.e., knowing exactly what data they are to collect and how they are going to format it. The issues of workforce shortage and lack of capacity in relevant fields are also shared between sectors in Brunei, where all sectors have stressed the importance of having trained personnel who are able to interpret collected climate change data and share the information contained within them.

Energy: Although the energy sector's needs are more focused on climate change mitigation rather than adaptation, it is acknowledged that to combat climate change, adaptation alone will not be fully effective. The moral responsibility for climate change is now focused on the oil industry (Grasso 2020). A global climate international tax on crude oil to finance programs to mitigate climate change is being studied (Anton 2020). Energy is a crucial sector for Brunei's economy and stability; mitigation measures in this sector is of high priority. The needs of this sector are geared toward achieving the aim of reducing anthropogenic emissions and the push for switching to renewable energy sources (EDPMO 2014). The needs of the sector include (i) Better records of methodologies used for past reported greenhouse gaseous (GHG) figures; (ii) Incentives to switch over to alternate energy sources (since alternate energy pricing tariffs are much higher even with government subsidies taken into account).

Water Resources: The relevant departments working on water resources are currently prioritizing water security, sanitation, flood prevention, coastal protection (from both erosion and rising sea levels), and the need to adopt smart development.

Smart development puts emphasis on allowing natural cycles to continue with minimal disturbance from said development in order to maintain unobstructed and optimum resource cycling. An example of this could be the planning and subsequent construction of a house that does not obstruct the local water flow, and hence the water returns to the water table. This helps to mitigate both flood risk and any negative effects on water cycling. Currently, the relevant water departments are not having much trouble meeting the country's demand for clean potable water (physical water availability is also not an issue). However, additional funding may be required in the future for water treatment depending on water use attitudes and demand. In terms of adaptive needs, the water sector requires (i) Safer and more efficient material transportation during dam construction; and (ii) New and effective ways of promoting responsible water use and awareness.

Agriculture: This sector primarily prioritizes self-sufficiency and food security. Secondary priorities involve the prevention of soil erosion and run-off associated with land cultivation. Despite having a set of dams intended purely for agricultural use, the main adaptive need of the agriculture sector is still heavily dependent on the water sector, i.e., the need for improved water security. Currently, most agricultural areas in Brunei are rainfed, and irrigation systems currently developed for the Brunei Muara district are expected to supply 1800 ha of agricultural areas (Page et al. 2009).

Health: The Public Health Sector deals with (and prioritizes) disease control, environmental health, food safety, and quality control and occupational health. The Public Health sector is divided into four divisions, each with its own set of responsibilities, functions, and goals. The collective goal of the Public Health Sector is to maintain excellent population health through effective disease control, environmental health management and food safety and quality control. The Public Health Sector has stressed the need (i) To include the Public Health Sector in the national climate change agenda; and (ii) For improved platforms for the drafting and approval of national actions plans (e.g., the National Environment and Health Action Plan).

Forestry: Brunei prioritizes the importance of sustainable development that keeps the country's forest intact. This is consistent with their mission that follows the developing, maintaining, and managing of forests for social, economic, and environmental benefits that will help the nation to prosper. For continued effective project implementation, and hence adaptation to climate change, the Forestry Sector has emphasized the need for an improved budget.

Biodiversity: The Heart of Borneo (HoB) Centre, which is the institutionalized body for the Heart of Borneo Initiative in Brunei, prioritizes the protection of the environment in order to ensure the continued unperturbed flow of other socioeconomic processes on a daily basis. The HoB Centre emphasizes the need to not only conserve Borneo's natural assets, but to also support economic growth in a sustainable manner. The main needs of the HoB Centre are more specialized research areas in order to maximize and fully exploit the contribution of the HoB conservation effort; improved practical, applicable, and relevant skill sets for their personnel.

2.3.2 *Potential Local Level Pilot*

Several vulnerable sectors requiring climate change adaptation measures for Brunei Darussalam have been identified. Of these sectors, water resources was identified as the most vulnerable sector. As there are currently several pilots already in place for this sector, it was proposed that the local level pilot focus on the next vulnerable sector, i.e., agriculture. Two of the main challenges for agriculture in Brunei are the development of irrigation systems to ensure water supply to agricultural areas, and the acid sulfate soils that require additional treatments to ensure their suitability for supporting agriculture (Grealish et al. 2008). The Brunei Government currently places high importance on agriculture, both for food security as well as economic diversification, and this emphasis therefore warrants a proposal for a local level pilot project.

The pilot aims to investigate the optimum growth conditions of climate resilient rice varieties and the degree at which they are able to resist the effects of climate change. Despite the wide range of traditional varieties with potential to resist climate change effects, the public demand for these crops tends to be restricted to rainfed areas. This is due to their low yields, long growth period, and relatively low photoperiod response, all of which limits cropping to once a year. Therefore, to compromise between climate change resilience and demand, rice will be used as the primary crop of interest for this pilot. The main objective is to determine under what specific conditions these rice varieties are best suited to grow in and why they are suited for post-climate change conditions. An additional outcome is also to quantify the relative performance of each variety of rice plants used in this pilot (all current reports are based on field observations). The rice varieties planned to be used in the pilot include Pusu (representing traditional variety in rainfed areas), BDR5 (representing improved variety in rainfed areas) and either MRQ76 or Titih (representing improved variety in irrigated areas).

The Department of Agriculture and Agrifood (DoAA), Ministry of Primary Resources is best suited to lead the proposed pilot since it is at the forefront of monitoring and shaping the agricultural land use of the country. The DoAA is committed to strengthening primary production sectors, including rice production (Department of Agriculture 2013) and the proposed pilot project would contribute greatly to the DoAA's efforts. The proposed pilot should be conducted in collaboration with UBD which currently places food security as one of its research thrusts, and would thus be in a position to provide academic and research expertise for this proposed pilot. The DoAA currently collaborates extensively with UBD on a crop-modeling project (Velde 2015). As such, this proposed pilot would complement current collaborative efforts between the DoAA and UBD.

The proposed study involves both fieldwork for the setting up of test plots (planting crops and selection of suitable sites), and lab-work for analyzing soil conditions and nutrient content. The pilot study can be carried out in any of the 66 Agricultural Development Areas (ADAs) present in Brunei, depending on the layout of the area and the practicality of setting up the varying conditions which will be investigated (pH range, soil nutrient, water content, etc.). To alleviate any burden imposed by the pilot onto the farmers, appropriate compensation should be granted. Compensation is usually approved by a designated panel under the DoAA when accompanied with the appropriate documentations and reasons, and after consultation with the respective ministerial bodies.

Each of these previously mentioned ADAs are divided into several 1-hectare plots depending on their local layout, and each of these are in turn further subdivided into several sub-plots whose dimensions are determined by the farmers cultivating the land. These smaller sub-plots function to allow the farmer to effectively contain the water in the sub-plots. The smaller sub-plots are separated by manufactured bunds that are compacted and sealed to prevent seepage. For the purpose of the pilot, the investigator in a few randomly selected areas isolated by bunds can manipulate the conditions being investigated. Replicating this setup at different ADAs can also improve the reliability of the pilot. The operational costs for the pilot was estimated by the DoAA to be around B$1,000 per hectare of farmland used per cropping season (costs include all manner of insecticides, herbicides, seeds, etc., but not labor force). Additional costs will be incurred if the investigation is destructive in nature, in which case additional compensation will need to be provided in order to cover the loss of income incurred by the farmers. The usual procedure practised by the DoAA for these types of compensation is to match the price value of the recorded yield from the previous season.

In terms of capacity needs, the study would require consultation from local and regional expertise (notably in the agricultural and/or paddy-growing business), including potential experts in ASEAN Member Countries with strengths in agriculture. In particular, Brunei Darussalam would like to have collaboration and skill sharing with Vietnam, Cambodia, and Myanmar in terms of dealing with acid sulphate soils and the varietal trials of high-yielding rice varieties. Other needs for this pilot include data on the predicted climate change effects for the region (most likely externally sourced), and greenhouse access (if certain environmental conditions cannot be simulated reliably in any of the existing ADAs).

The expected outcome of this pilot is the optimization of agricultural land output, i.e., enabling farmers to produce more yields without the conversion of other types of land into agricultural lands. The findings from this study could be used to improve yield and hence further improve the country's self-sufficiency status, which is consistent with the Agriculture Sector's priorities. At the same time, the Energy and Industry Department at the Prime Minister's Office could also benefit from extra yields in the form of a boosted local industry. The water sector will also stand to gain since agriculture is deeply tied to water resources. The income generated by a boost in agriculture and industry could be re-invested into water projects to help improve water frameworks in agriculture. There is also potential for future international collaboration with other countries that are using these particular crop species as well as for technology and skills transfer. All of these possibilities make for good opportunities for co-financing. The budget will be spent on operating costs, the hiring of skilled labor, equipment for both field and lab studies, and for the provision of paddy seeds.

2.4 Concluding Remarks

Climate change measures in Brunei Darussalam are still in its early stages. Limited knowledge, capacity and resources are currently getting in the way of a truly effective climate change adaptation plan. However, the progression of the relevant national documents of the country on climate change adaptation reveals that it is definitely heading in the right direction: the cross-sectoral nature of the Brunei Darussalam Strategic National Action Plan (SNAP) that enables inter-disciplinary efforts is definitely a key factor that needs to be implemented to ensure a successful climate change action plan. With time, as the expertise and technology of Brunei improves, the country will establish a dedicated and effective climate change adaptation plan all the way from the drafting process through to the monitoring phase.

Currently, Brunei Darussalam does not have a dedicated climate change policy, and measures for adaptation and mitigation tends to be scattered around within the framework of different projects that each department undertakes. The first step towards establishing a dedicated policy on climate change for Brunei Darussalam commenced as Brunei drafted its National Communications and Intended Nationally Determined Contributions under the UNFCCC. After the establishment of these documents, it is expected that more projects and pilots, which are designed specifically to address climate change mitigation and adaptation, will be implemented. A local level pilot focused on agriculture is proposed, to be located in any of the 66 suitable Agricultural Development Areas of Brunei. The pilot will involve both fieldwork for the setting up of test plots and lab-work for analyzing soil conditions and nutrient content.

Acknowledgements We thank the Department of Environment, Parks and Recreation, Ministry of Development, for their guidance and support. We are grateful to the following stakeholders for providing inputs: Energy Department at the Prime Minister's Office; Department of Water Services, Ministry of Development; Forestry Department, Ministry of Primary Resources and Tourism; Department of Agriculture and Agrifsood, Ministry of Primary Resources and Tourism; Brunei Heart of Borneo Centre, Ministry of Primary Resources and Tourism; Environmental Health Services, Ministry of Health; and Universiti Brunei Darussalam. The authors acknowledge the ASEAN-India Green Fund for supporting the Project on "Enhancing Local Level Climate Change Adaptation in Southeast Asia" (Phase 1). This chapter draws partly on findings of the Project, specifically the local level needs assessment and scoping exercise conducted in 2015.

References

Abidin PHZ, Ahmad PSW (2006) Supplement to government gazette part II. Jabatan Percetakan Kerajaan, Brunei Darussalam

Airudin F (2014) Tutong flood mitigation projects. https://www.bt.com.bn/news-national/2014/05/25/tutong-flood-mitigation-projects. Accessed 25 May 2017

Anon (2015a) Green building initiatives. https://www.pwd.gov.bn/index.php?option=com_content&view=article&id=179&Itemid=395. Accessed 25 May 2017

Anon (2015b) Heart of Borneo Spine proposal to be reviewed. https://www.bt.com.bn/news-national/2014/08/26/heart-borneo-spine-proposal-be-reviewed. Accessed 26 Aug 2017

Anton A (2020) Taxing crude oil: a financing alternative to mitigate climate change? Energy Policy 136 (2020):111031

ASEAN Inter-Parliamentary Assembly (2012) Brunei Darussalam country report on the disaster management. In: The 4th AIPA Caucus. Bangkok, 30 April–2 May 2012

Azaraimy HH (2015) Sultanate's own green building rating tool soon. https://borneobulletin.com.bn/sultanates-own-green-building-rating-tool-soon/. Accessed 24 May 2017

Baines PG (2011) Patterns of decadal climate variability and their impact on global rainfall. Procedia Environ Sci 6:70–87

Bengtsson L, Hodges KI (2006) Storm tracks and climate change. J Clim 19(15):3518–3543

BruDirect (2004) Brunei Darussalam the abode of peace. https://www.bruneidirecthys.net/about_brunei/geography.html. Accessed 25 June 2015

Brunei Darussalam Meteorological Department (2015) Climate. https://www.bruneiweather.com.bn/climate. Accessed 8 June 2016

Brunei Darussalam (2007) Wild fauna and flora order. http://www.agc.gov.bn/AGC%20Images/LOB/Order/U-Z/Wild%20Fauna%20and%20Flora%20Order,%202007%20(S%2077).pdf. Accessed 3 October 2021

Brunei Darussalam's Second National Communication (2017) Energy and industry department, the prime minister's office. Unpublished internal document, Brunei Darussalam

Brunei Times (2011) National council on climate change set up with His Majesty's consent. http://www.bt.com.bn/news-national/2011/02/14/national-council-climate-change-set-hismajestys-consent. Accessed 8 June 2016

Chan H (2015) The heart of Borneo corridor project: ensuring healthy watershed corridors for Borneo. In: international heart of Borneo (HoB) conference on bridging HoB landscapes and beyond through healthy watershed corridors, Kota Kinabalu, 11–12 Nov 2015

Cheong CG (2010) Bruneians receiving power from the sun. The Brunei Times

Dale VH, Joyce LA, McNulty S, Neilson RP, Ayres MP, Flannigan MD, Hanson PJ, Irland LC, Lugo AE, Peterson CJ, Simberloff D, Swanson FJ, Stocks BJ, Wotton BM (2001) Climate change and forest disturbances. Bioscience 51(9):723–734

Department of Agriculture (2013) Roles and functions. http://www.agriculture.gov.bn/DOA2009/contents/vision.html. Accessed 20 Nov 2015

Disaster Management Order (2016) Constitution of Brunei Darussalam (order under article 83(3)). Brunei Darussalam

EDPMO (2014) Energy white paper on Bandar Seri Begawan, Brunei Darussalam. Energy department prime minister's office

Forestry Department (2010) Fourth national report for Brunei Darussalam. Publication, Ministry of industry and primary resources

Grasso M (2020) Towards a broader climate ethics: confronting the oil industry with morally relevant facts. Energy Res Soc Sci 62(2020):101383. https://doi.org/10.1016/j.erss.2019.101383

Grealish GJ, Fitzpatrick RW, Ringrose-Voase AJ, Hicks W (2008) Brunei: summary of acid sulfate soils. In: Fitzpatrick RW, Shand P (eds) Inland acid sulfate soil systems across Australia. CRC LEME open fi le report no. 249 (Thematic volume) CRC LEME, Perth, Australia, pp 301–309

Hasan DSNABPA, Ratnayake U, Shams S, Nayan ZBH, Rahman EKA (2019) Prediction of climate change in Brunei Darussalam using statistical downscaling model. Theoret Appl Climatol 133(1–2):343–360

Heart of Borneo (2012) About the HoB initiative. https://www.hobgreeneconomy.org/en/about/. Accessed 28 May 2015

HoB Trilateral Conference (2015) The heart of borneo R&D trilateral conference: towards three countries one conservation vision. 13–14 April 2015. Press Release. Accessed 3 October 2021 http://www.agriculture.gov.bn/SiteCollectionDocuments/Press%20Release%202015/April/13.04.2015_hob%20r%20%20d%20conference%2013%2014%20april%202015_%20press%20release.pdfhttp://www.agriculture.gov.bn/SiteCollectionDocuments/Press%20Release%202015/April/13.04.2015_hob%20r%20%20d%20conference%2013%2014%20april%202015_%20press%20release.pdf

iCUBE (2014) Gastropod thermal biology and climate change workshop. https://www.icubeconsortium.org/index.php/component/content/article/91-workshops/136-gastropod-thermal-biology-and-climate-change-in-the-tropics. Accessed 28 Dec 2019

INFORM (2017) The inter-agency standing committee task team for preparedness and resilience and the European commission. https://www.inform-index.org. Accessed 13 Dec 2018

Johnson JA (2012) Assessing the impact of climate change in Borneo. World Wildlife Fund, Washington DC

Le Quéré C, Raupach MR, Canadell JG, Marland G (2009) Trends in the sources and sinks of carbon dioxide. Nat Geosci 2:831–836

Loo YY, Billa L, Singh A (2014) Effect of climate change on seasonal monsoon in Asia and its impact on the variability of monsoon rainfall in Southeast Asia. Geosci Front 6(6):817–823

Manton MJ, Della-Marta PM, Haylock MR, Hennessy KJ, Nicholls N, Chambers LE, Collins DA, Daw G, Finet A, Gunawan D, Inape K, Isobe H, Kestin TS, Lefale P, Leyu CH, Lwin T, Maitrepierre L, Ouprasitwong N, Page CM, Pahalad J, Plummer N, Salinger MJ, Suppiah R, Tran VL, Trewin B, Tibig I, Yee D (2001) Trends in extreme daily rainfall and temperature in Southeast Asia and the South Pacific: 1961–1998. Int J Climatol 21(3):269–284

Masli U (2011) Stick energy labels on appliances. The Brunei Times

Michener WK, Blood ER, Bildstein KL, Brinson MM, Gardner LR (1997) Climate change, hurricanes and tropical storms, and rising sea level in coastal wetlands. Ecol Appl 7(3):770–801

Ndah AB, Odihi JO (2017) A systematic study of disaster risk in Brunei Darussalam and options for vulnerability-based disaster risk reduction. Inte J Disaster Risk Sci 8(2):208–223

Nicholls RJ, Hoozemans FMJ, Marchand M (1999) Increasing flood risk and wetland losses due to global sea-level rise: regional and global analyses. Glob Environ Chang 9(Supplement 1):S69–S87

Othman A (2016) El-Nino weather to continue until March. Borneo Bulletin

Page S, Hoscilo A, Langner A, Tansey K, Siegert F, Limin S, Rieley J (2009) Tropical peatland fires in Southeast Asia. In: Tropical fire ecology. Springer Praxis Books, Springer, Berlin, Heidelberg

Public Works Department (2015) Green building initiatives. http://www.pwd.gov.bn/index.php?option=com_content&view=article&id=179&Itemid=395. Accessed 29 Sep 2015

Sadikin S (2014) Jail, caning for illegal logging. The Brunei Times, Brunei Darussalam

Shams S, Zahran ESMM, Said SNM, Hui KJ, Lee NDHMF, Hasim NHH (2019) Risk assessment for forest fire in Brunei Darussalam. MATEC Web of Conf 258:05033. https://doi.org/10.1051/matecconf/201925805033

Silvestre G, Matdanan HJH, Sharifuddin PYH, De Silva MWRN, Chua TE. (1992) The coastal resources of Brunei Darussalam: status utilization and management. In: Proceedings of the ASEAN/US CRMP workshop on coastal resources management in Brunei Darussalam, 30 April–1 May 1991, pp 1–38

The Guardian (2014) Indonesia moves to stop forest fire pollution as haze grips Singapore. http://www.theguardian.com/environment/2014/sep/16/indonesia-forest-fire-pollution-haze-singapore-palm-oil. Accessed 25 June 2015

UNISDR (2007) Hyogo framework for action 2005–2015: Building the resilience of nations and communities to disasters. United Nations

Velde BV (2015) Forest news update: Q&A on fires and haze in Southeast Asia. https://blog.cifor.org/23000/forests-news-update-qa-on-fires-and-haze-in-southeast-asia?utm_source=Fire+research+landing+page&utm_medium=bellow+story&utm_campaign=Fire+research+Package#.VZDiphuqqko. Accessed 25 June 2015

Wich SM, Streubig J, Refish A, Wilting S, Kramer-Schadt EM (2015) The future of the Bornean Orangutan: impacts of change in land cover and climate. Publication, UNEP/GRASP

Wong A (2015) Retention basins to help mitigate flooding. http://www.bt.com.bn/legco/2015/03/18/retention-basins-help-mitigateflooding. Accessed 18 March 2015

Chapter 3
Climate Change Adaptation in Cambodia

Sothun Nop

Abstract Cambodia, a tropical country with high average temperatures and two distinct seasons, a dry and rainy season. Despite the fact that climate change impacts have occurred in the communities, the adaptation mechanism of the local government and community remains very limited. A local level pilot is proposed in Kampong Speu, which is a vulnerable province with a high incidence of poverty. Farmers are expected to face the impact of drought and local communities have limited understanding of related root causes and the potential disasters. The expected loss and damage is still under investigation but anticipated social problems include loss of livelihood, high rates of school dropout, and other aspects of social vulnerability.

Keywords Climate change adaptation · Climate impacts · Vulnerability · Disaster risks · Cambodia

3.1 Introduction

With a total area of 181,035 km^2 that make up 25 provinces, Cambodia in mainland Southeast Asia shares its borders with three countries: Thailand, Laos PDR, and Vietnam (Fig. 3.1.) It is one of the countries that the famous Mekong River flows through as it makes it way from the north into the Mekong Delta of Vietnam in the south. During the rainy season, the Mekong river also feeds the Tonle Sap Lake, which is in the northwest region of Cambodia (World Bank 2011).

The northern mountains, flat midlands, and the south-western coastal region make up the Cambodian mainlandkey priority sectors include agriculture (Nop 2015). The Dangrek, Cardamom, and Elephant mountain ranges can be found in the north and northeast region where the main crops are rubber, cassava, cashews, among others. The flat midlands around Tonle Sap Lake is a major rice production area.

S. Nop (✉)
Faculty of Development Studies, Royal University of Phnom Penh, Phnom Penh, Cambodia
e-mail: nop.sothun@rupp.edu.kh; sothun.nop@gmail.com

School of Science, The University of New South Wales, Canberra, Australia

J. J. Pereira et al. (eds.), *Climate Change Adaptation in Southeast Asia*, Disaster Risk Reduction, https://doi.org/10.1007/978-981-16-6088-7_3

Fig. 3.1 Cambodia is surrounded by Thailand, Lao PDR and Vietnam. *Source* Nationasonline.org

The southwest coastline is about 435 km long and hosts agro-industrial crops particularly oil palm and pepper. Mangrove forests are prevalent in some areas along the coast (Nguyen and Shaw 2010; RGC 2001; UNDP 2013b; World Bank 2011). Deforestation has resulted in reduction of significant carbon stocks from mangroves; the maintenance and restoration of these areas is an effective strategy for climate change mitigation and adaptation (Sharma et al. 2020).

Being a tropical country, Cambodia has high average temperatures with a dry and rainy season. The rainy season is from May to October where about 75% to 80% of the annual rainfall is brought by south-westerly winds. The dry season occurs from November to April with average temperatures ranging from 27 to 40 °C. The months between October and January are the coolest and most pleasant (Nguyen and Shaw 2010; RGC 2001).

Cambodia is a lower middle income country with a total population of 15.33 million in 2014 (World Bank 2015). Predominantly agrarian, more than 80% of the total population are located in rural areas. About 52% of the population live in the central plains and about 30% are around the Tonle Sap Lake. The country is still recovering from the civil war under the Pol Pot regime between 1975 and 1979. The war destroyed much of the major physical infrastructure making the social and economic restoration process a major challenge (Nguyen and Shaw 2010; RGC 2001; UNDP 2013b).

The economic growth of Cambodia has been strong over the past decade (UNDP 2013a). In the 2000s, the annual GDP growth rate fluctuated between 5 and 13%. The Gross Domestic Product (GDP) is USD 10,354 million per annum relying primarily on agriculture and services, which contribute 21 and 42% of the GDP, respectively (World Bank 2011). Both these sectors employ 15.9 and 26% of the country's labor force. Whilst contribution of the industrial sector to the GDP has doubled since 1993, farming and fisheries remain the primary contributing sector of the country (World Bank 2011). The Cambodian economy is estimated to have grown by 7.6% in 2013 (UNDP 2013a). The Ministry of Economics and Finance (MEF) reports that the annual GDP per capita is USD 1,036 compared with around USD 200 in 1992. Over the last ten years, rapid economic growth and national renovation were fuelled by substantial natural resources; this resulted in significant changes in land use. In 2004, there were large acreages of forest which were cleared to make way for agricultural and barren land (Fig. 3.2).

A majority of Cambodians live in rural areas and approximately 45.9% of the total population live in multidimensional poverty (as determined by using factors beyond income-based lists) and depend heavily on climate-sensitive resources (UNDP 2013a). As one of the three least developed nations of the ASEAN Member States, Cambodia is very vulnerable to climate change impacts due to limited adaptive capacity (UNDP 2013c; Yusuf and Francisco 2010). The country is exposed to climate hazards such as floods and droughts as a consequence of the over-exploitation of natural resources such as forests, land, water, and minerals in the quest for economic development. Such environmental issues have brought about many challenges for building community resilience in the context of climate change and

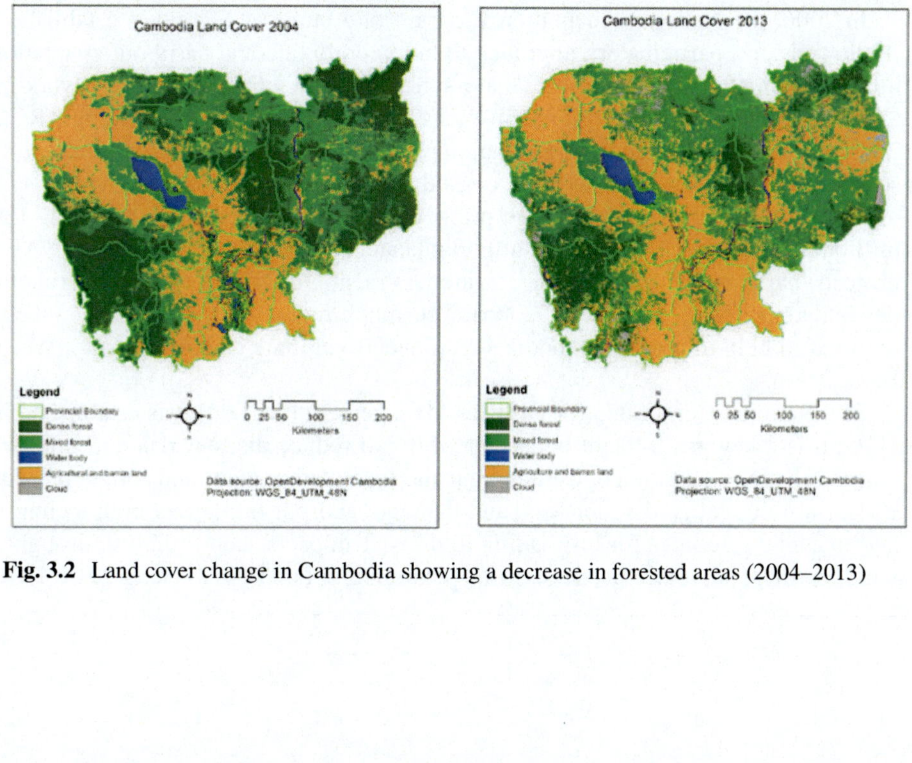

Fig. 3.2 Land cover change in Cambodia showing a decrease in forested areas (2004–2013)

for Cambodian development as a whole (Lay 2011; Sreng 2013). Public investment in the expansion and rehabilitation of irrigation and drainage infrastructure is critical to support agriculture production and rural households to adapt to conditions of floods and droughts (Nguyen et al. 2020).

3.2 National Perspective

3.2.1 National Arrangements and Institutions

The Royal Government of Cambodia (RGC) has developed its climate change adaptation policy and guidelines to minimize the impacts of climate change, and to help improve community adaptive capacity; it did this by increasing the roles and responsibilities of relevant ministries and institutions from national to grassroots levels (CDRI 2012; RGC 2006; Sreng 2013). A framework for implementation of the policy and guidelines has also been established. Like its ASEAN counterparts, the government also ratified the policy of the United Nations Framework Convention on Climate Change (UNFCCC) in 1995 and the Kyoto Protocol in 2002. In 2003, a Climate Change Office was set up which later became a department in 2009. In 2006, the National Climate Change Committee (NCCC) was formed under Sub-Decree N° 35. This national committee is made up of the Prime Minister as president, and 19 ministries and relevant agencies; a minister from the Ministry of Environment (MoE) chairs the NCCC and other working groups and task forces make up a technical team (RGC 2006).

In 2006, the National Adaptation Program of Action (NAPA) was established. It allowed for a participatory approach to help coordinate and carry out adaptation initiatives. NAPA also aimed to create synergies with other institutions working toward similar goals of development so as to help with capacity building. (RGC 2006). Priority NAPA projects were put in place in the country for important sectors such as agriculture, water resources, coastal zones, and human health (Heng 2011); of a total of 39 projects, 20 are high priority, with a combined budget of USD 130 million (World Bank 2011). Three important categories are the focus in these NAPA projects: capacity building/training, awareness raising/education, and infrastructure development (Heng 2011; RGC 2006). Yet, much more support and assistance is required to help improve Cambodia's resilience to climate change impacts (World Bank 2011).

The National Committee for Disaster Management (NCDM) was established in 1995; it is a key government body that works to reduce disaster risk and improve climate change adaptation by strategizing and formulating plans and action, provide early warning, gauge disaster loss, and damage, and put into effect both technical and financial resources for responding to disaster impacts such as floods, droughts, wind storms, and vector-borne diseases (Chhinh and Poch 2012; Heng 2014; RGC

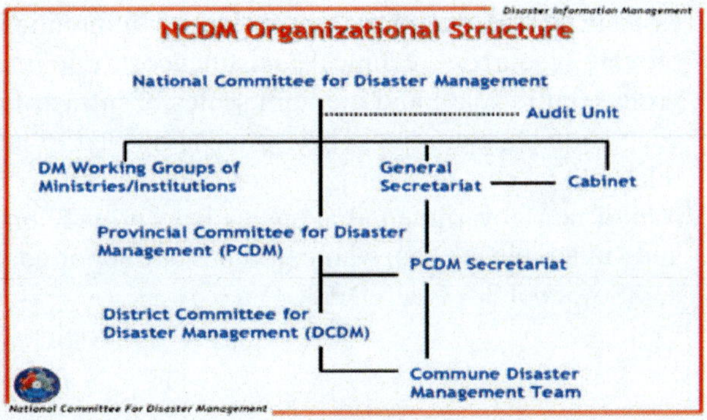

Fig. 3.3 NCDM organizational structure. *Source* Sreng (2013)

Undated). The NCDM works closely with its sub-national branches (provincial, district, and commune disaster management committees) (Fig. 3.3).

Plans for adaptation to climate change have been included in national policies and strategic guidelines. For example, the Rectangular Strategies, the National Strategic Development Plan (NSDP) 2006–2010, and the NSDP-Update 2014–2018 were established to reduce poverty and improve national development. Also, relevant ministers, development partners, and local communities are working together to better promote the effectiveness of the implementation of climate change policies and guidelines (Lay 2011; RGC 2006; Sreng 2013).

3.2.2 *Implementation Mechanisms*

There are several government bodies and development agencies from national to local levels involved in carrying out adaptation policies and guidelines. The NCCC is the lead body; it is responsible for disseminating policies to relevant stakeholders, and it mobilizes resources and collective action for implementing the priority projects. The NCC is chaired by the Minister of Environment (MoE). The MoE at the national level has limited power in handling droughts and floods as it does not have top-down line authority with the grassroot level (Chhinh and Poch 2012). The MoE solved this by working closely with NCDM which had its agents from the national to local level.

In addition, there are other relevant state agencies that contribute to dealing with the issue of disaster and climate change. The Ministry of Economic and Finance (MEF) provides financial resources and the Ministry of Planning (MoP) provides technical support for climate actions (Nguyen and Shaw 2010; Saito 2013). The Ministry of Agriculture, Forestry and Fishery (MAFF) and the Ministry of Water Resource and Meteorology (MoWRAM) provide technical support and capacity building at the sub-national level, specifically the provincial departments and district offices (Chhinh and Poch 2012).

Many non-government organizations (NGOs), development partners, and private sector entities are also involved in this national issue. Their primary functions are to assist in education and training of local communities so that they are more prepared, provide resources (technical and finance) to upgrade the livelihood of vulnerable groups, build small and medium scales of infrastructure through community-based adaptation programmes. They are also needed to influence government bodies and relevant stakeholders for pro-poor policies and social justice. Beside this, another role of non-governmental agencies is to provide emergency response to the needy and vulnerable groups, who could not be supported by the government (D'Agostino and Sovacool 2011; UNDP 2011).

3.2.3 Vulnerable Sectors and Regions

Climate change is increasing the temperature and changing the rainfall patterns in Cambodia; resulting in extreme events such as floods and droughts (Nang et al. 2014). The baseline for rainfall and temperature has a spatial variation across the country (Fig. 3.4). Since 1960, the mean annual temperatures have increased by 0.8 °C, at a rate of about 0.18 °C per decade (World Bank 2011). During the dry season (December–January–February and March–April–May), the rate of temperature increase is higher, about 0.20–0.23 °C per decade. The rate is slower during the wet seasons (June–July–August and September–October–November), about 0.13–0.16 °C per decade. The temperature in Cambodia is expected to increase between 0.3 and 0.6 °C by 2025 (Baran et al. 2009). A similar projection of temperature increase between 0.7 and 2.7 °C by 2060s, and between 1.4 and 4.3 °C by 2090s has been

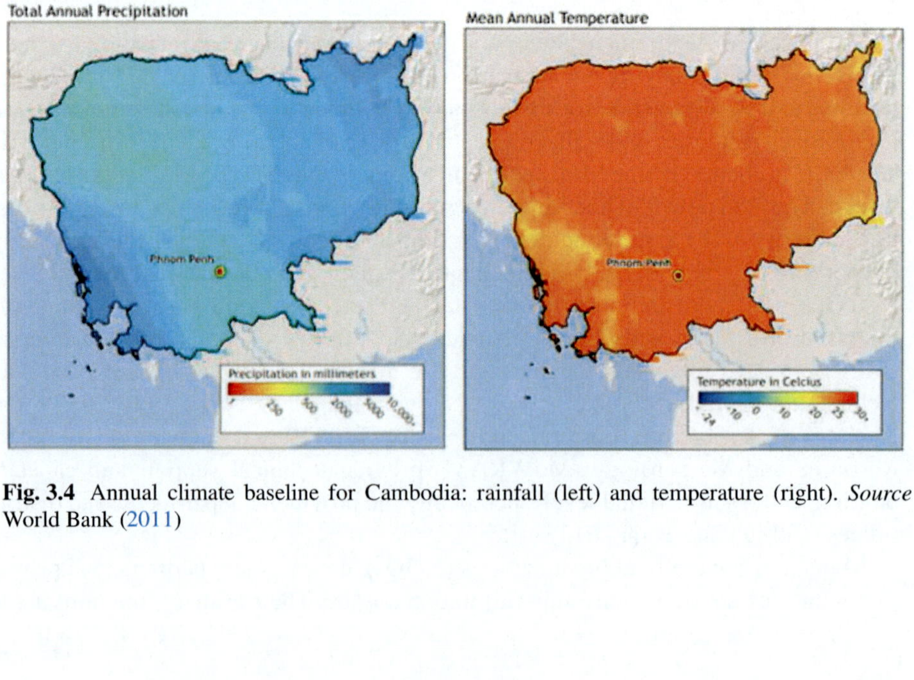

Fig. 3.4 Annual climate baseline for Cambodia: rainfall (left) and temperature (right). *Source* World Bank (2011)

reported (McSweeney et al. 2008). There is concern that agricultural productivity will be affected by the increase in temperature (RGC 2013).

With every increase of 10 °C in growing-season minimum temperature in the dry season, a 10% decrease in rice grain yield is expected (International Rice Research Institute cited in RGC 2013).

Since the 1960s, the annual total rainfall has remained consistent in Cambodia. The rainfall is expected to increase; to the magnitude of 1- and 5-day rainfalls of up to 54 mm and 84 mm, respectively, by the 2090s (McSweeney et al. 2008). While the amount of rainfall remains the same, changes are expected in rainfall pattern (Diepart 2014). Prior to the 1980s, the rainfall season began in May, peaked in July, and ended in November. In the 1980s, less rainfall was observed in May but the amount continued increasing to its peak in September, and declined in November. The variation in rainfall caused crop failure due to water shortage. The RGC (2013) reported that the change of rainfall pattern brings about negative effects on the production of rubber and coffee in Cambodia.

There is much concern even though there is no clear scientific evidence that the increase in sea level is affecting the country's coastal zones; many areas already suffer from storm surges, high tides, beach erosion, and sea-water intrusion. The World Bank (2011) recognizes that sea-level rise is a significant threat to marine coastal areas. Sea level in the region is projected to increase between 21 and 52 cm by the 2090s. This threatens low-lying areas that comprise housing, tourism facilities, fishery infrastructure, and mangrove forests. A one-meter rise of sea level is expected to inundate about 44 km^2 of coastline in Koh Kong; significantly increasing the risk of flooding as well as impacting freshwater quality in the area (Unknown 2014). RGC (2013) further suggested that sea-level rise would make the Cambodian coastline more vulnerable to natural hazards, particularly typhoons and strong winds that can affect tourism potential as well as coastal settlements.

In the last ten years, the country has been struck by more severe floods and droughts. According to NCDM cited by Heng (2014), there has been a rising trend of flood intensity between the years of 2000 and 2013. Twenty of its 25 provinces experienced severe floods in 2013 alone. Floods mostly affect the provinces along the Mekong River including Kandal, Kampong Cham, Kratie, Prey Veng, Stung Treng, Svay Rieng, and Takeo (World Bank 2011). Consequently, lives and crops were lost and people suffered from vector-borne diseases. The same study also indicated that the trend of drought intensity has also increased in the same period; fourteen of 25 provinces of Cambodia experienced severe droughts in 2012 resulting in crop failure and land degradation.

NCDM (2013), moreover, confirmed that between 1996 and 2013, 2,050 lives were lost due to disasters. Flood damaged about 1.7 million hectares (accounting for 67% of total losses) of paddy fields, whereas drought caused the devastation of around 0.77 million hectares (31%). A total of 45,372 houses, 194 hospitals as well as schools, roads, and irrigation systems were also demolished. The average annual economic loss of due to floods is estimated at USD 16.1 million while USD 7 million is accounted to drought (UNISDR and World Bank 2010).

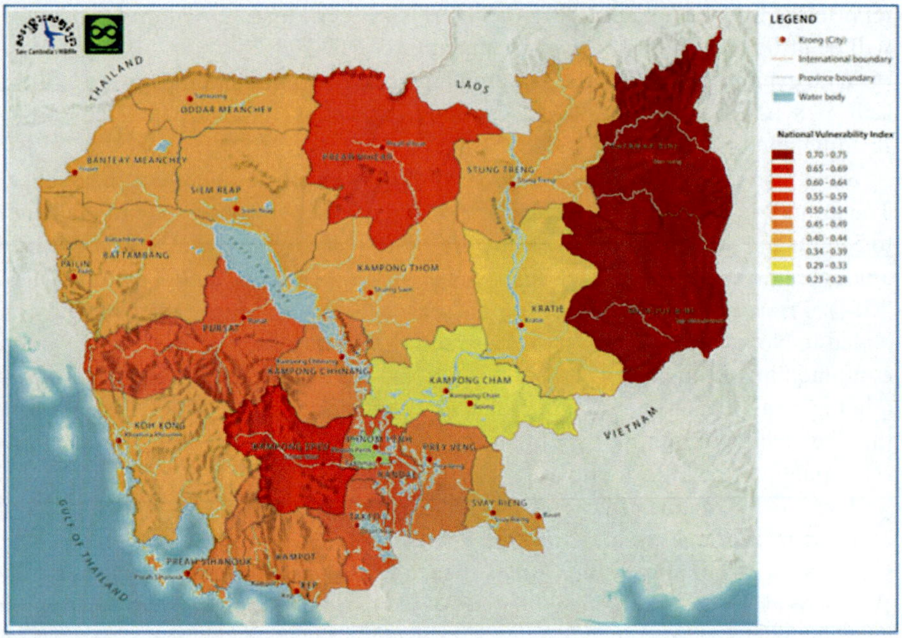

Fig. 3.5 National vulnerability index. *Source* Open Development Cambodia (2015)

As mentioned in the literature, Cambodia is currently encountering many climate -related hazards including floods, droughts, windstorms, soil erosion, salt-water intrusion, and diseases. With greater frequency and intensity of these climate hazards, the key sectors including agriculture, health, water resource lives, and biodiversity are at high risk. Consequently, lives of communities whose livelihoods rely on these climate-sensitive sectors could be severely affected (CDRI 2012; Nguyen and Shaw 2010; RGC 2013; Sreng 2013; World Bank 2011).

The vulnerability index is based on the criteria that the region has been frequently exposed and potentially affected by climate-related hazards (Fig. 3.5). The adaptive capacity of local community and institution was also included in the key criteria for classification.

3.3 Adaptation Priorities

3.3.1 Key Sectors

Adaptation needs and priorities were determined through a combination of literature review and stakeholder consultation. In order to improve adaptive capacity and reduce climate change impacts, key priority sectors include agriculture, human health, and water resources, among others (World Bank 2011; RGC 2013).

Agriculture: Since a large proportion of Cambodian people rely on agriculture, improving this sector will help ensure food security as well as maintain their livelihoods, which greatly contribute to improving the adaptive capacity, especially in the local communities. According to World Bank (2011), the agricultural sector can be improved in several ways. Existing support for research and extension services has to be enhanced in the sector. In addition, the efficiency in use of water for harvesting and small-scale irrigation, need improvement. The capacity of farmers in accessing weather data, diversification of activities is also a key aspect for improvement. There is also a need to upgrade dikes for flood protection and establish seed banks to benefit the community.

Health: As average temperatures and changes in rainfall patterns occur, the mechanisms of water- and vector-borne disease transmission in the country are being affected. As 45% of its population does not have access to public health facilities, it is not surprising that Cambodia has one of the highest rates of malaria deaths in Asia (World Bank 2011). When NAPA surveys were carried out in seven provinces, it was discovered that malaria was most transmitted in agricultural fields and forests are most common in; this was rife during the wet months. NAPA contended that the situation is made worse because of ignorance about malaria. The prevalence of malaria, dengue fever, and water-borne diseases like cholera, dysentery is expected to be higher; this combined with food shortage will gravely affect the lives and livelihoods of a high percentage of the country's population. With more rainfall, mosquito habitats and populations will thrive, further exacerbating the situation. As total eradication of malaria is next to impossible, other alternatives have to be considered. The health sector needs to be enhanced: (i) implementing programs to educate the community are vital as people need to be aware of the importance of personal hygiene as well as making the environment a healthy living area; (ii) a surveillance system to detect malaria has to be set up, together with better methods for vector control for health workers and communities; (iii) capacity building through effective and relevant training programs and enhancing health-related services at local levels; and (iv) supporting health research and community health services.

Water Resources: The predominant sources of water in Cambodia are the Mekong River and its tributaries, while underground water supplies serve both domestic and irrigation. Despite Cambodia having many sources of water, access to water all year round remains a problem to many Cambodian people; and underground water in some places has arsenic contamination, which is harmful for health. About 81% of households surveyed have reported water shortages (RGC 2006). To improve water resource management mechanisms and deal with the issue of water shortage, the following priorities have been suggested: (i) Carry out assessment of water resources and create an inventory of water quality and quantity of surface and underground water in time and space, so as to enable optimum use of existing water resources; (ii) Bring in improved methods of water conservation, storage, and use, as well as build small check dams and teach ways to harvest rainwater to supplement domestic and irrigation supplies; (iii) Set up community-involved watershed management and conservation programs to restore degraded watersheds; (iv) Introduce methods for flood protection, disaster prevention, and maintenance of flood control structures; and (v) Improve provincial water transportation arteries (reducing silt, dredging), particularly along the upper Mekong waterways in the provinces.

Other Aspects: While stakeholder consultation tends to agree with the key priorities suggested in the literature, the other key sectors like transport, tourism, soil quality, forestry, and coastal zone protection need to be taken into account as these will also contribute to improving the livelihoods of the communities, balancing the ecosystem and prove beneficial for promoting sustainable national development in the long run. Stakeholder consultation also called for disaster risk reduction to be integrated with climate change adaptation as such a holistic approach can help to build a concrete foundation for the key actor, especially community, to confront extremes climate risks. In the meantime, integrating and introducing both these aspects to other important sectors such as transport, tourism, forestry, soil, and coastal erosion are needed in the effort to keep all relevant actors informed and organized to combat climate change events.

3.3.2 Potential Local Level Pilot

A local level pilot was proposed to integrate loss and damage into disaster risk reduction and climate change adaptation using the case of the Peary Meanchey Dam in Borsedth District, Kampong Speu Province. Kampong Speu is one of the provinces of Cambodia located about 48 km west of the municipality of Phnom Penh. The total area of the province is 7,017 km^2 and it is administratively divided into one town (Krong Chbar Morn) and 7 districts. According to Yusuf and Francisco (2010), Kampong Speu is the third most vulnerable province after Mondulkiri and Ratatakiri (Fig. 3.5). Kampong Speu was also assessed as the most affected by droughts because it experienced them in high frequency, and they caused huge damage to paddy fields (Chhinh 2013; Chhinh and Poch 2012).

Borsedth is one of the districts of Kampong Speu Province located in the southern region of Cambodia, which shares its border with Kampot and Takeo Province to the south and two other districts, Samrongtong to the north and Kong Pisei to the northeast. The district consists of flat farmland and supports extensive agriculture and a large population. Borsedth is the second largest district of Kampong Speu Province with a total population of 108,6648 (52.9% female) in 1998. This district is subdivided into 15 communes and 218 villages (NIS 2000).

Since Borsedth is one of the most vulnerable districts of Kampong Speu, the dam project, which was designed for climate change adaptation was built there. The dam is essential for agriculture (rice production) and water supply for families in the district. The proposed pilot will benefit the communities both short and long term. It will immediately assist communities to seek alternatives and options for preserving water resources for their agricultural production with their engagement in growing organic crops for domestic consumption and selling. For the medium- and long-term benefits, the quality of their agricultural land areas can be promoted because when farmers access water resources, they will grow more crops; and this will help conserve soil quality in the long run.

The Faculty of Development Studies (FDS) at the Royal University of Phnom Penh (RUPP) is expected to play a key role as a lead institution for the project implementation. Key relevant state agencies including Department of Water Resource and Meteorology (DWRAM), Department of Agriculture, Forestry and Fishery (DAFF) of Kampong Speu Province will also collaborate. The DWRAM will be in charge of providing information and support regarding water resource use and management in the proposed study areas, while DAFF will provide all relevant information in agricultural techniques, support, and production. Moreover, local authority and community-based organizations/associations for community representation will be participating throughout the pilot project. It is expected that through the involvement of local authority and communities, there will be a greater sense of ownership on the project which will help to maintain the project outcomes in the long term.

The potential outcomes of the pilot include: (i) Enhanced knowledge of local community on climate change impacts; (ii) Improved water resource use and management skills in the area; (iii) Reduced impacts of climate change diseases and disasters; and (iv) Improved diversified livelihood activities of the local community. The effectiveness of water use and management will also be improved as relevant stakeholders, especially local community, become more cautious and aware of the importance of water resources. In addition, this project will contribute to enhance networking between all relevant stakeholders and create more spaces for future collaboration.

Despite the fact that climate change impacts have occurred in the communities, the adaptation mechanism of the local government and community remains very limited. In this regard, the capacity of communities and local government could be enhanced by offering key knowledge and skills such as (i) Understanding on impacts of climate-related hazards; (ii) Framework for adaptation action; (iii) Water resource management; (iv) Leadership and financial management skills; (v) Agricultural technique development; (vi) Proposal Writing; and (vii) Project Planning and implementation.

To implement this pilot project, it is expected that core research fund of RUPP will be applied for. Besides this, possible research and development funding from the Ministry of Water Resource and Meteorology (MoWRAM), Ministry of Agriculture, Forestry and Fishery (MAFF), and other potential donors or funding agencies will also be assessed. This pilot project is expected to be implemented through the support of partners on climate change adaptation within the country. Those key institutions include the Ministry of Environment (MoE), Ministry of Water Resource and Meteorology (MoWRAM), Ministry of Agriculture, Forestry and Fishery (MAFF), and Ministry of Land Management, Urban Planning and Construction (MLMUPC), who will be sharing the data or reports related to the topics. The proposed pilot project could also be implemented through collaboration with the research partner institutions within ASEAN Member States, including Malaysia, Vietnam, or Laos PDR, where similar issues exist. In addition, the practical techniques for implementing the project will be consulted among the experts from those member countries. The results and successful lessons learned from this pilot project will be shared among this network.

3.4 Concluding Remarks

Cambodia is guided by a framework of policies and guidelines to reduce climate risks and enhance adaptive capacity, under the coordination of the National Climate Change Committee (NCCC) established in 2006, which is chaired by the Minister of Environment (MoE). The National Adaptation Program of Action (NAPA) was formulated in 2006. The roles and responsibilities of relevant government agencies at various levels have increased to tackle climate change. The National Committee for Disaster Management (NCDM) established in 1995 is the key government body that works with sub-national entities to reduce disaster risk and improve climate change adaptation, at the provincial, district, and communal levels. The MoE has limited linkages at the communal level to tackle hazards such as floods and droughts, and works in collaboration with NCDM to tackle disaster events associated with climate change.

Priority sectors for climate change adaptation are agriculture, human health, and water resources. The potential for a decrease in rice grain yield, increased prevalence of malaria, dengue fever, and water-borne diseases as well as water shortage due to extreme climate events affect millions of Cambodians. Other key aspects of importance to adaptation include integration between disaster risk reduction and climate change adaptation as well as the protection of coastal areas. Floods and droughts have increased in the last decades, causing loss of lives, damage to crops, and serious vector-borne diseases. Such incidences are expected to increase with climate change. Sea level rise is projected to make the coastline more vulnerable to natural hazards, particularly typhoons and strong winds, which could affect tourism and coastal settlements. Enhanced capacity of key actors as well as awareness building and coordination of all stakeholder groups is critical to strengthen climate change in Cambodia. A local level pilot is proposed in the Kampong Speu Province, which has a high level of poverty and is most affected by drought in the country, impacting rice production; and to integrate loss and damage into disaster risk reduction and climate change adaptation.

Acknowledgements The author acknowledges the ASEAN-India Green Fund for supporting the Project on "Enhancing Local Level Climate Change Adaptation in Southeast Asia" (Phase 1). This chapter draws partly on findings of the Project, specifically the local level needs assessment and scoping exercise conducted in 2015.

References

Baran E, Schwartz N, Kura Y (2009) Climate change and fisheries: vulnerability and adaptation in Cambodia. World Fish Center, Penang

CDRI (2012) Challenge of rural livelihoods in the context of climate change. Cambodia Dev Rev 16(3):1–5

Chhinh N (2013) Climate change adaptation for Cambodian agriculture: a policy review climate change vulnerability assessment Kampong Speu Province. Royal University of Phnom Penh, Phnom Penh

Chhinh N, Poch B (2012) Climate change impacts on agriculture and vulnerability as expected poverty of Kampong Speu Province, Cambodia. Int J Environ Rural Dev 3(2):28–37

D'Agostino AL, Sovacool BK (2011) Sewing climate-resilient seeds: implementing climate change adaptation best practices in rural Cambodia. Mitig Adapt Strateg Glob Change 16(6):699–720. https://doi.org/10.1007/s11027-011-9289-7

Diepart JC (2014) Learning for socio-ecological resilience: insights from Cambodia's rural communities. Development research forum. Phnom Penh, Cambodia

Heng CT (2011) Status of NAPA implementation in Cambodia. Second national forum on climate change Cambodia. Cambodia

Heng LA (2014) Country report of Cambodia disaster managment Kobe, JAPAN. Asian Disaster Reduction Centre (ADRC), Japan

Lay K (2011) National dialogue on mainstreaming DRR into development practices and trends. Disaster risk reduction forum, Cambodia

McSweeney C, New M, Lizcano G (2008) UNDP climate change country profiles. School of geography and the environment of the University of Oxford. UNDP, Oxford

Nang P, Sam S, Lonn P, Ouch C (2014) Strengthening Adaptation Capacity of Rural People in the Main Agro-ecological Zone in Cambodia, (Vol. 05). Cambodian Development Resource Institute, Phnom Penh

NCDM (2013) Cambodia disaster loss and damage analysis report 1996–2013. National committee for disaster management/United Nation development program. Phnom Penh

Nguyen H, Shaw R (2010) Climate change adaptation and disaster risk reduction in Cambodia. Commun Environ Disaster Risk Manag 5:59–79

Nguyen TT Nguyen TT Grotea U (2020) Multiple shocks and households' choice of coping strategies in rural Cambodia. Ecol Econ 167(2020):106442. https://doi.org/10.1016/j.ecolecon.2019.106442

NIS (2000) General population census of Cambodia, 1998. National Institute of Statistics, Cambodia

Open Development Cambodia (2015) Climate change vulnerability in Cambodia (2010). Date Accessed, 10 December 2016. Retrieved from https://data.opendevelopmentcambodia.net/en/dataset/map_vulnerability_index

RGC (2001) Draft second five-year socioeconomic development plan 2001–2005. Ministry of planning. Phnom Penh, Cambodia

RGC (2006) National adaptation programme of action to climate change (NAPA). Royal Government of Cambodia, Phnom Penh

RGC (2013) Cambodia climate change strategic plan 2014–2023. National Climate Change Committee, Phnom Penh, Cambodia

RGC (Undated) Cambodia's intended nationally determined contribution

Saito N (2013) Mainstreaming climate change adaptation in least developed countries in South and Southeast Asia. Mitig Adapt Strateg Glob Change 18(6):825–849. https://doi.org/10.1007/s11027-012-9392-4

Sharma S, MacKenzie RA, Tieng T, Soben K, Tulyasuwan N, Resanond A, Blate G, Litton CM (2020) The impacts of degradation, deforestation and restoration on mangrove ecosystem carbon stocks across Cambodia. Sci Total Environ. 706:135416. https://doi.org/10.1016/j.scitotenv.2019.135416

Nop S (2015) Towards active community participation in implementing climate change adaptation policy (CCAP) in Cambodia. MPRA Working paper no. 71656, Munich Personal RePEc Archive, Munich, Germany. https://mpra.ub.uni-muenchen.de/71656/

Sreng S (2013) Climate change issues in Cambodian coastal area related to water. Climate change issue and urban flood management. Daegu, Korea. http://www.unescap.org/esd/suds/egm/2013/urban-flood-management/index.asp. Accessed 12 June 2015

UNDP (2011) Promoting climate resilient water management and agricultural practices in rural Cambodia (NAPA Follow Up) annual project report 2010. United Nation development program. Phnom Penh

UNDP (2013a) About Cambodia retrieved 15 Jan 2015. https://www.kh.undp.org/content/cambodia/en/home/countryinfo/. Accessed 18 June 2015

UNDP (2013b) About Cambodia retrieved 22 Aug 2013. https://www.kh.undp.org/content/cambodia/en/home/countryinfo/. Accessed 18 June 2015

UNDP (2013c) Case study in Cambodia community based adaptation: two examples from rural affected communities. United Nation development program. Phnom Penh, Cambodia

UNISDR & World Bank (2010) Synthesis report on ten ASEAN countries disaster risks assessment. United Nation office for disaster risk reduction & The World Bank

Unknown (2014) Supplementary appendix M—CAM GMS BCC-PPCR climate change impact modeling and vulnerability assessments for Koh Kong and Mondulkiri Provinces in Cambodia

World Bank (2011) Vulnerability, risk reduction and adaptation to climate change. The World Bank Group, Cambodia

World Bank (2015) Cambodia population retrieved 20 Feb 2015. https://data.worldbank.org/country/cambodia. Accessed 25 December 2015

Yusuf AA, Francisco H (2010) Hotspots! mapping climate change vulnerability in Southeast Asia. Singapore economy and environment program for Southeast Asia (EEPSEA)

Chapter 4
Climate Change Adaptation in Indonesia

Laksmi Rachmawati, Andini Desita Ekaputri, Luh Kitty Katherina, and Heru Santoso

Abstract A large proportion of Indonesia is prone to climate change. Vulnerabilities portrayed at the national and regional levels may not link well with those portrayed at the local level. Most studies at the national and regional levels successfully identify vulnerable hot spots. However, they fail to understand the causes of social vulnerability. It is important to understand the causes of social vulnerability as each individual and community have different levels of access to resources, coping capacity, and recovery capacity. A comprehensive vulnerability assessment is necessary especially at the local level to better understand causes of social vulnerability, which are often context specific. Two areas are proposed for enhancing local level climate change adaptation in Indonesia, Lombok Island in West Nusa Tenggara Province and Probolinggo in East Java Province.

Keywords Climate change adaptation · Climate impacts · Vulnerability · Disaster risks · Indonesia

4.1 Introduction

Efforts on climate change adaptation are always behind mitigation. Global attention to climate change started during the Rio Earth Summit in 1992. National leaders from 172 countries gathered in Rio de Janeiro, Brazil from 3–14 June 1992 to discuss the importance of finding alternative energy to replace fossil fuels that are linked to climate change (http://www.un.org/geninfo/bp/enviro.html). The Summit agreed on the climate change convention with the purpose to stabilize greenhouse gas concentrations in the atmosphere at a level that would prevent dangerous anthropogenic interference with the climate system. The convention then led to the establishment of Kyoto Protocol that provides mechanisms to stabilize greenhouses gases (GHGs).

L. Rachmawati (✉) · A. D. Ekaputri · L. K. Katherina · H. Santoso
National Research and Innovation Agency (BRIN) Indonesia,
Jakarta, Indonesia

H. Santoso
e-mail: heru.santoso@brin.go.id

J. J. Pereira et al. (eds.), *Climate Change Adaptation in Southeast Asia*,
Disaster Risk Reduction, https://doi.org/10.1007/978-981-16-6088-7_4

The decision was adopted during the third session of the Conference of Parties (COP) in Kyoto in 1997. Impacts of climate change and adaptation were addressed during the first session of COP in Berlin in 1995.

However, the first milestone in implementing adaptation was in 2001 during the 7th session of COP in Marrakesh (http://unfccc.int/focus/adaptation/items/6999.php). Parties agreed to establish a Least Developed Countries (LDC) Work Programme. The work programme focused on the development of national climate change mechanism and capacity building through National Adaptation Programmes of Action for the LDCs. Later, the Nairobi Work Programme (NWP) was adopted in 2006 during the 12th session of COP in Nairobi to increase the knowledge and capacity on impacts, vulnerability, and adaptation in relation to climate change. The Cancun Adaptation Framework (CAF) was established during the 16th session of COP in Cancun in 2010 to facilitate developing countries, in particular the LDCs, to formulate and implement their National Adaptation Plans (NAPs) and to enable developed countries to support adaptation in developing countries.

Climate change impacts some countries more severely, and Indonesia as an archipelagic country faces many challenges. Vulnerability and risk assessments are critically needed as a requirement for planning adaptation strategy and action. Vulnerability assessments (VAs) are crucial for well-planned adaptation efforts. National level vulnerability assessments in Indonesia started in 2009–2010 when the Indonesian Climate Change Sectoral Roadmap (ICCSR) was published by the National Development Planning Board, BAPPENAS (Badan Perencana Pembangunan Nasional) to guide vulnerability assessments in Indonesia. After 2009, there have been several vulnerability assessments in some areas and sectors that were conducted by sectoral ministries, local governments, and other parties including non-government organizations (NGOs).

This chapter briefly highlights findings from vulnerability assessments conducted in Indonesia. The report commences with a general overview of policies on climate change in Indonesia. This is followed by a review of vulnerability assessments in Indonesia at multiple scales. The final section is a proposal for enhancing local level climate change adaptation.

4.2 National Perspective

4.2.1 National Arrangements and Institutions

Indonesia contributes actively in combating the impacts of climate change both at global and national level. Indonesia ratified the climate change convention through Act No. 6 in 1994, two years after the global treaty on climate change convention signed during the Rio Earth Summit in 1992. Indonesia also ratified the Kyoto Protocol in 2004 through Act No. 17, almost seven years after the Kyoto Protocol was established and adopted during the 3rd session of COP in Kyoto, in 1997. The significance to address climate change in national development was stated in the National 2005–2025 Long-Term Development Plan (RPJP 2005–2025). This was described in more detail in the 2004–2009 National Medium-Term Development Plan (RPJMN 2004–2009) which identified potential risks of climate change on some development targets. However, RPJMN 2004–2009 did not describe sectoral and cross-sectoral action plans in addressing climate change.

In 2007, Indonesia hosted the 13th session of COP in Bali. Indonesia launched the National Action Plan addressing climate change (NAP/RAN-PI—*Rencana Aksi Nasional Perubahan Iklim*) during the COP 13. This COP was a stepping-stone to increasing awareness of climate change actions, both mitigation and adaptation, in relevant sectors (MoE 2007). The RAN-PI marked multi-sectoral efforts in implementing climate change actions. Soon after this COP, the President of the Republic of Indonesia established the National Climate Change Council (NCCC/DNPI—*Dewan Nasional Perubahan Iklim*) through the Presidential Regulation No. 46/2008. The purpose of DNPI was to coordinate Indonesia's activities in addressing climate change and to strengthen Indonesia's position in international forums on climate change.

RAN-PI is the first national document that describes sector and cross-sector actions for climate change mitigation and adaptation. Following the RAN-PI, Ministry of Environment (MoE), National Development Planning Agency (BAPPENAS) and the National Climate Change Council released a number of documents, including policies. They outlined strategies and plans of action to address climate change as well as enable actions to strengthen institutional capacity. The documents include the following:

- Indonesia Country Report: Climate Variability and Climate Changes, and Their Implications, published by Ministry of Environment and Ministry of Public Works, December 2007. The objective is to describe climate variability and climate change projections, including sea level rise in Indonesia, and their impacts on development sectors.
- National Development Planning: Indonesia Responses to Climate Change ("Yellow Book"), published by Ministry of National Development Planning (BAPPENAS) in June 2008. The objective is to define mitigation actions and priority sectors for adaptation. It includes funding for mitigation and adaptation, and the list of climate change-related projects of prioritized sectors and non-prioritized sectors.
- Indonesia Climate Change Sectoral Road Map (ICCSR), published by BAPPENAS in March 2010. The objective is to create a detailed road map for each sector to implement actions in responding to climate change. ICCSR provides input for the development of the 2010–2014 Medium-Term Development Plan (RPJMN 2010–2014) and its consecutive periods in 30 years.
- Adaptation Science and Policy Study (DNPI, 2010). The objective is to review the state of science and policies on climate change adaptation in Indonesia. It supports prioritized adaptation actions in ICCSR by identifying gaps between the

needs and availability of technologies to help adaptation actions, including the required fund.

- National Action Plan for Greenhouse Gas Emission Reduction (RAN GRK – Rencana Aksi Nasional Penurunan Emisi Gas Rumah Kaca), endorsed by BAPPENAS in December 2010. The objective is to design programmes and actions to reduce greenhouse gas emission in five main sectors, namely, forest and peat land, agriculture, industry, energy and transportation, and waste. The designed programmes and actions are to be implemented at national and local scales within the framework of sustainable development. The document was then legally supported by a Presidential Regulation No. 61/2011 regarding National Action Plan on GHG Emission Reduction.
- National Action Plan for Climate Change Adaptation (RAN-API—*Rencana Aksi Nasional Adaptasi Perubahan Iklim*), published by DNPI (DNPI, 2011). The objective is to provide adaptation action plans for each sector based on their needs, planned programmes, and their implementation, for better coordination between sectors. This DNPI version of RAN-API also serves as input in the development of the BAPPENAS version of RAN-API.
- Indonesia Adaptation Strategy: Improving Capacity to Adapt (BAPPENAS, 2011). The objective is to update knowledge on Indonesian climate change projections, including sea level rise in Indonesia and the impacts on sectors, and strategies to adapt.
- Strategy to mainstream adaptation into National Development Plan: integration framework(BAPPENAS, 2012). The objective is to provide an integration framework for mainstreaming adaptation at the national and local levels, and sectors.
- Development of Data and Information System of Climate Change Adaptation, published by DNPI in September 2012 (Impron et al. 2012). The objective is to provide information on adaptation activities in Indonesia and to review their conformity in the context of climate change adaptation.
- National Action Plan for Climate Change Adaptation (RAN-API). A Synthesis Report was published by BAPPENAS in November 2012. Updated synthesis report 2013. The complete (full) RAN-API was published in February 2014. The objective of RAN-API is to establish a national adaptation plan of actions. The main purpose is to implement sustainable development with high resiliency to climate change. The plan was developed through coordination and participation of relevant government sectors, civil societies, and private sectors.

Indonesia submitted its Second National Communications in 2011 (MoE 2010). The Third National Communication was submitted in 2018 (MoEF 2018). The National Communications documents the country's state in the greenhouse gases inventory and its efforts on climate change mitigation and adaptation. Of all the national documents, there are four major ones that provide the basis for deriving implementation actions as follows:

- RAN-PI (MoE 2007). It can be considered as a milestone in addressing the climate change issue in the context of development in general. It was the first time sectors acknowledged climate change as a development issue instead of an environmental issue only.

- "Yellow Book" (BAPPENAS 2008). It serves as a basis for the development of implementing actions in response to climate change. The Yellow Book was also a reason for the establishment of the Indonesia Climate Change Trust Fund (ICCTF) through National Development Planning Ministry Decision No. 144/2009. ICCTF serves as a funding mechanism in supporting the implementation of climate change mitigation and adaptation in Indonesia. It was legally strengthened through a Presidential Regulation No. 80/2011 regarding establishment, structure, and operation of a trust fund.
- ICCSR (BAPPENAS 2010a). This document provides detailed implementing actions for each priority sector in responding to climate change, and also input for the Medium-Term Development Plan (RPJMN 2010–2014) and consecutive Long-Term National Development Plan. It is considered as the consequence of the previously published "Yellow Book".
- RAN GRK (BAPPENAS 2010b). This document contains implementation programmes and actions to reduce greenhouse gas emission in five main sectors in RAN-API. It was then legally endorsed through Presidential Regulation No. 61/2011 on RAN GRK. Soon after this, a Presidential Regulation No. 71/2011 regarding the National GHG Inventory System was released in order to provide the legal basis for producing regular information on the level, status, and trend of greenhouse gases emission change and absorption in Indonesia. It was then followed by a Presidential Decree No. 19/2010 on the preparation for the development of an institution for Reducing Emission from Forest Degradation and Deforestation Plus (REDD +), Presidential Instruction No. 10/2011 on forest moratorium that includes the development of REDD + schemes, and a Presidential Decree No. 25/2011 on National Task Force for REDD + that has been renewed by a Presidential Decree No. 5/2013.

4.2.2 National Level Vulnerability Assessments

In 2010, BAPPENAS launched a national document called the Indonesia Climate Change Sectoral Roadmap (ICCSR). This ICCSR project involved many experts and multiple stakeholders from various development sectors and regions in Indonesia in order to develop vulnerability assessments for some important sectors and regions.

The report aims at providing scientific-based risk assessment necessary for national development plan in terms of medium- and long-term planning. The ICCSR summary report (p. 6) states explicitly that the objectives of the report are to develop national goals, sectoral targets, milestones, and priorities for actions regarding adaptation and mitigation to climate change for all development sectors.

The increasing number of climate hazards become the main rationale in developing this report. With the changing climate, there would be an increase in surface air temperature, precipitation trend, sea level rise, and extreme climate events. With the national assessment scale, the results of the assessment are presented at a macro level represented by "big" islands such as Sumatra, Java-Bali, Kalimantan, Sulawesi, Nusa Tenggara, Maluku, and Papua.

There were several volumes or series of reports, preceded by two volumes on a scientific basis: (i) analysis and projection of temperature and rainfall, and (ii) analysis and projection of sea level rises and extreme events. The other volumes comprise information on risk assessment and strategies to adapt in several development sectors. The sectors discussed in this series describe high priority sectors for adaptation: i.e. water resources, marine and fisheries, and agriculture sectors (Table 4.1). High priority for mitigation includes the forestry, energy, industry, transportation, and waste management sectors.

In 2009, Yusuf and Francisco (2009) published "Climate Change Vulnerability Mapping for Southeast Asia". Their study was based on indicators that used secondary data (Fig. 4.1). The study covered seven countries in the Southeast Asian region, consisting of 530 sub-national areas. The indicators used are as follows:

- Exposure: (i) historical record of climate hazard, and (ii) climate-hazard maps for five climate hazards (tropical cyclones, floods, landslides, droughts, and sea level rise);
- Sensitivity: (i) population density, and (ii) ecological sensitivity to climate-hazard exposure.
- Adaptive capacity: (i) socioeconomic factors, (ii) technology, and (iii) infrastructure.

Climate hazards were an important parameter for exposure in this vulnerability assessment. The assessment delineated Java-Bali, Nusa Tenggara, Papua, and a small part of Sumatera as being more vulnerable compared to other areas in Indonesia.

Considering the importance of vulnerability assessment, the Ministry of Environment (KLH), which has since been restructured as the Ministry of Environment and Forestry (KLHK), established a Data Inventory System for Vulnerability Indices, SIDIK (*Sistem Inventori Data Indeks Kerentanan*) for measuring vulnerability in Indonesia at the provincial, district or city and village levels (KLH 2009). Some of the indicators used are based on secondary data from publications such as Village Potentials PODES (*Potensi Desa*), published sectoral data, climate data from the National Agency for Meteorology, Climatology, and Geophysics, BMKG (*Badan Meteorologi, Klimatologi dan Geofisika*) as well as climate projection data. The SIDIK comprises 28 indicators of which 14 indicators were for adaptive capacity and the other 14 for exposures and sensitivity. The SIDIK is still in the process of development and new indicators will be added as work progresses on the system over time (DGCC 2015). The information in SIDIK can be used by multi-stakeholders including the Ministry of Environment at the national level, the board of environment in districts and cities, and other users such as the private sector.

Table 4.1 The summary of sector assessment from the Indonesia Climate Change Sectoral Roadmap (ICCSR) Synthesis Report (2010)

Sector	Hazards	Projection
Water	Water shortage	– Java-Bali have already undergone water deficit (in water balance) – Java-Bali and Sumatra will face severe water shortage for the period of 2020–2030 *Extremely high risk*: northern and southern part of West Java, middle and southern central Java and East Java, North Sumatra, West Sumatra, Bengkulu and Lampung, West Nusa Tenggara and South Sulawesi – Other areas will suffer from a gradual decrease of water supply due to temperature increase and rainfall changes
	Floods	*Extremely high risk*: areas along major rivers, particularly in downstream areas of Java, Eastern Sumatra, most parts of Western-Southern and Eastern Kalimantan, Eastern Sulawesi and Southern Papua
	Droughts	*Extremely high risk*: Central Java, Sumatra and Nusa Tenggara
Marine and fisheries	Coastal inundation	Java Island, particularly in the North Coastal area
	Sea surface temperature	In 2030, temperature increase by 1–2 °C could increase coral bleaching
	Extreme events	– Increased frequency of El Nino and La Nina (from an average of every 3–7 years to every 2–6 years) – Increasing extreme weather events such as storm surges, tropical cyclones and high waves
Agriculture	Food production	– Area of paddy fields will decrease due to sea level rise – Decreasing production due to temperature increase and precipitation decrease – Extreme weather events also adversely affect food production

(continued)

Table 4.1 (continued)

Sector	Hazards	Projection
	Plantation production	– Decreasing production in coffee, cacao, and palm oil
Health	Vector-borne infectious diseases	– Increasing morbidity and mortality due to climate disasters – Increasing malnutrition because of decreasing food production which adversely affects food supplies – Increasing diseases triggered by changing temperatures, air pollution, water congenital disease, food and congenital disease vector and rodents
	Diarrhoeal diseases	

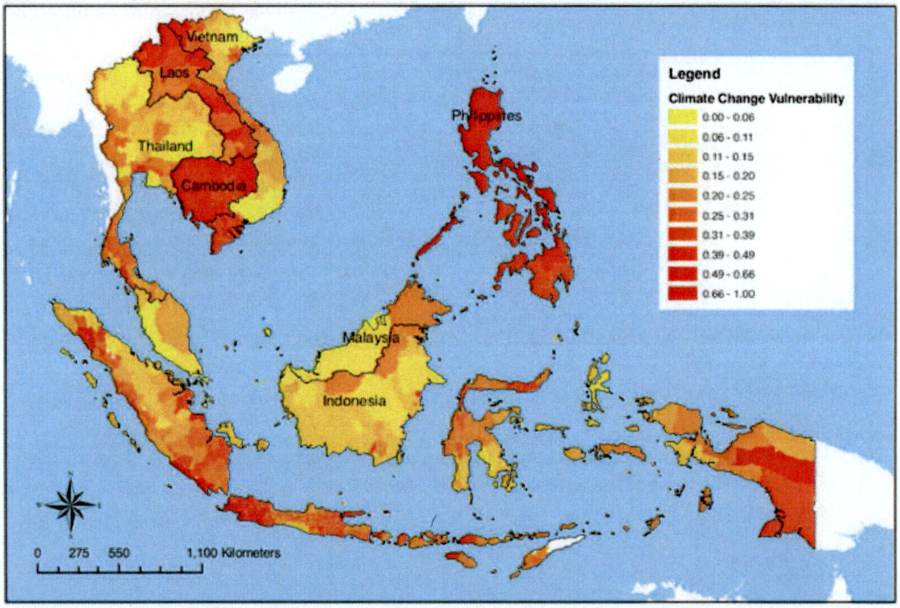

Fig. 4.1 Climate change vulnerability map of Southeast Asia. *Source* Yusuf and Francisco (2009)

4.2.3 Regional Level Vulnerability Assessments

Regional VAs covers several areas of Indonesia (Fig. 4.2). The VAs have been conducted by multiple parties (Table 4.2). The VAs at the regional level were conducted by government institutions such as the Ministry of Environment (KRAPI) and National Council of Climate Change (DNPI). Donor agencies such as GIZ, Mercy Corps, and JICA have also conducted VAs (Fig. 4.3). The VAs are necessary before any adaptation intervention is taken and in several areas a sectoral approach

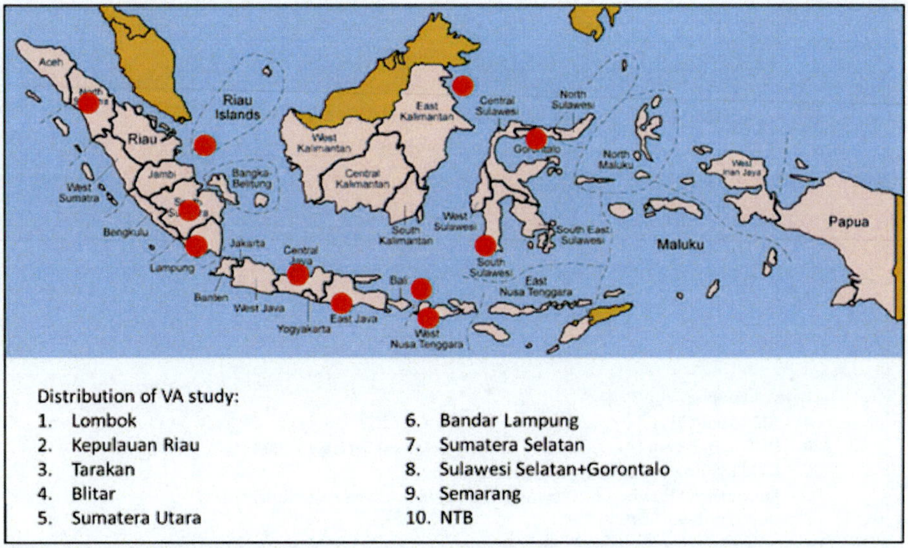

Fig. 4.2 Distribution of vulnerability assessment studies in Indonesia

Table 4.2 Summary of vulnerability assessments conducted by multiple parties in Indonesia

No	Location	Year	Conducted by	Sectors				
				Coastal	Water	Health	Agriculture	Disaster
1	Lombok	2009	GTZ, WWF	v	v		v	
2	Bandar Lampung	2010	Mercy Corps					v
3	Semarang	2010	Mercy Corps		v			v
4	Tarakan	2012	KRAPI	v	v	v		
5	South Sumatera	2012	KRAPI	v	v	v	v	
6	Malang Raya	2012	KRAPI					
7	North Sumatera	2012	JICA				v	
		2013	DNPI				v	
8	Kepulauan Riau	2013	DNPI					v
9	Blitar	2013						v
10	South Sulawesi and Gorontalo	2013	DNPI			v	v	
11	Bali	2013	JICA				v	

Fig. 4.3 Distribution of vulnerability assessment studies conducted by donors

is taken (Fig. 4.4). The VAs were conducted at the island level in Bali and Lombok. However, Bali also represents a provincial level VA. The others VAs such as those for Kepulauan Riau, Sumatera Utara, Sulawesi Selatan, and Gorontalo were conducted at the provincial level. Some VAs such as those for Semarang, Blitar and Bandar Lampung were conducted at the municipality level by Mercy Corps of the Asian

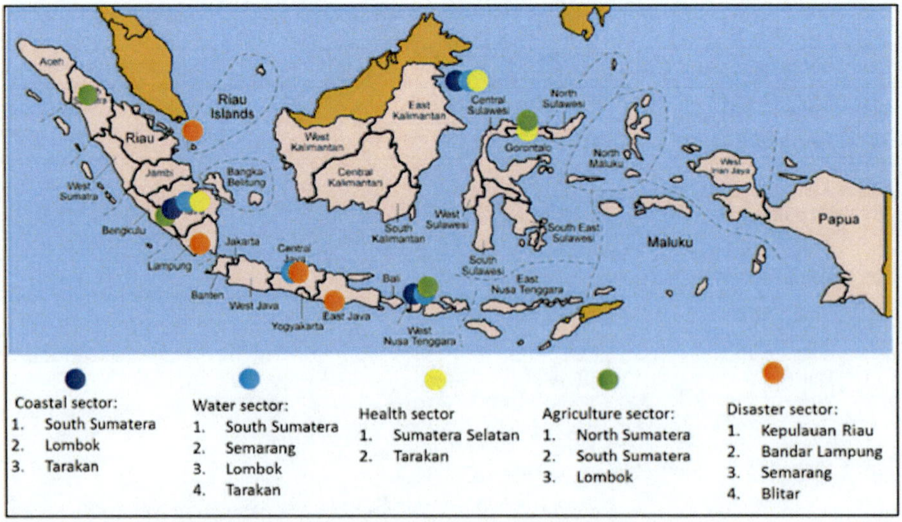

Fig. 4.4 Distribution of vulnerability assessment studies by sector

Table 4.3 Indicators used for the water sector in Tarakan and South Sumatra

Hazards	Exposure	Sensitivity	Adaptive capacity
Floods Landslides	– Population density – Land use	Role of infrastructure	Income
Water shortage	– Population water demand – Land use water demand	Sources of water	Income

Cities Climate Change Resilience Network (ACCCRN) funded by the Rockefeller Foundation.

Climate change risk assessments have been conducted in three areas i.e. Lombok, South Sumatera, and the small island of Tarakan. The studies come from the Risk Assessment and Climate Change Adaptation, KRAPI (*Kajian Risiko dan Adaptasi Perubahan Iklim*) projects. The KRAPI projects used the following indicators for assessing risk: topography, slope, land use, population density and numbers, infrastructure as well as critical facilities. In Lombok, the assessment used an additional indicator, which is the welfare level. In general, the findings from these three assessments highlighted the increasing risk of climate hazards, which would adversely affect the population, particularly in certain cities. These studies also highlighted sea level rise, storm surges, El Nino, and high waves as potential climate hazards. The population and infrastructure in coastal areas of South Sumatera are more vulnerable compared with inland areas. Thus, Palembang City and Musi Banyuasin District are the most vulnerable compared with other districts. The findings from a study in the small island of Tarakan show that its coastal areas have a high level of vulnerability as reflected by most of the indicators used. The Eastern part of the Tarakan Coastal area is less vulnerable compared with other coastal areas. In addition, the capital of Tarakan City is the most vulnerable. The study in Lombok shows that the city area (Mataram City) is more vulnerable compared with the suburban districts.

Vulnerability assessments for the water sector were conducted in Lombok, Tarakan, South Sumatra, and Semarang City. Highlighted climate related hazards are floods, water shortage, and landslides (Table 4.3). The findings show that in South Sumatra, the area that has an existing risk of floods and landslides is Palembang City. However, in 2030, the flooded area is projected to expand to The West and The South of Palembang City. Water shortage is also projected to expand from the Banyuasin Disrict in 2010 to The West and The South in 2030, which includes Ogan Komering Ulu, Ogan Komering Ulu Timur, Ogan Komering Ilir, Lubuk Linggau and Prabumulih.

The vulnerability assessment in Tarakan used additional indicators in the adaptive capacity component to that in South Sumatra. These include information on landslide hazards, piped water, and water shortage from government programmes. Water-related hazards are projected to increase in these areas with climate change. The high-risk area for floods is in the centre of the city. The area in the southwest has the highest risk for landslides compared with other areas. Water shortage also will become a serious problem in 2030, albeit at lower magnitude than floods and landslides. Areas with increased magnitude of water shortage will expand to all coastal areas.

In Lombok, among the three water problems, water shortage will become the most severe in 2030. The most vulnerable area will be in the centre of the island, linked to serious problems in Mataram City. The study shows that floods are not a serious problem for the island at the present and in the future. Using different methodology and indicators, Mercy Corps conducted a vulnerability assessment in Semarang on some water-related hazards such as floods, tidal inundation, sea level rise, droughts, and landslides. The assessment shows that the coastal areas and eastern parts of Semarang have high vulnerability.

With respect to the health sector, the coverage was on areas with dengue, malaria, and diarrhoea epidemics (Table 4.4). The vulnerability assessment results show a significant expansion of the areas that have dengue epidemic from Palembang City in 2008 to more than seven areas in 2030. These are Muara Enim, Musi Banyuasin, Ogan Komering Ulu Selatan, Ogan Komering Ulu Timur, Prabumulih, Pagar Alam and Lubuk Linggau. For malaria, the 2008 baseline records the areas of Ogan Komering Ulu Timur and Ogan Komering Ilir and Musi Rawas. However in 2030, this will expand to almost the whole of South Sumatra except Palembang City. A similar situation projected for diarrhoea, where from the 2008 baseline areas that comprises Palembang, Musi Banyuasin and Banyuasin, will expand to include Ogan Komering Ulu, Muara Enim, Musi Banyuasin, Banyuasin, South Ogan Komering Ulu, and East Ogan Komering Ulu. The rising trends from baseline year to projection year also occur in Tarakan.

Table 4.4 Indicators used for the health sector in Tarakan and South Sumatra

	South Sumatra	Tarakan
Dengue fever	E: population S: source of water supply; urban population density; mobility of people (travellers and seasonal migrant workers) AC: provision of health facilities and accessibility to health facilities	– Amount of population – Population density – Amount of water supply – Provision of health facilities
Malaria	E: population S: distance from mosquito breeding site; type of housing; type of profession; AC: availability of mangrove area; provision of health facilities and accessibility to health facilities	– population near breeding site – house near breeding site – non-permanent housing – health facilities
Diarrhoea	E: population S: Household sanitation facility; source of water supply; prolonged flood area; proportion of sensitive age people (infant and old people) AC: immunization; provision of health facilities and accessibility to health facilities	– number of population – house without toilet – pipe water coverage – health facilities

Note E: exposure, S: sensitivity and AC: adaptive capacity

Indicators for vulnerability assessments in the agriculture sector are used to measure vulnerability from exposure component, sensitivity, and adaptive capacity. In terms of exposures, they are areas of land for agriculture and numbers of farmers. For sensitivity, they are agriculture land types, topography, and farmers' income. For adaptive capacity component, they are irrigation infrastructure, education level of farmers, and agriculture sector income to regional domestic product. With the projected climate change, there will be an expansion of areas that will have problems in agricultural production. In South Sumatera, the baseline areas currently suffering from agricultural production are Musi Banyuasin, Ogan Komering Ulu Timur, Banyuasin, Ogan Ilir and Ogan Komering Ilir. The areas suffering will increase to include Musi Rawas, Muara Enim, Empat Lawang and Prabumulih. The severely decreased precipitation projected for almost all areas of Lombok is expected to lower agriculture production from the entire area.

4.2.4 Local Level Vulnerability Assessments

The study related to vulnerability assessment at the local level is limited. The United States Agency for International Development (USAID) and the Ministry of Marine and Fisheries (MMF) have jointly conducted a vulnerability assessment at the local level. Using a method referred to as I-catch (Indonesia Climate Adaptation Tool for Coastal Habitats), the study covered West Nusa Tenggara and Southeast Sulawesi. The objective was to understand the vulnerability of people who live in coastal areas and facilitate adaptation and mitigation actions based on the findings. The framework is based on the understanding that the impact of climate change in coastal areas is manifested by increasing numbers of threats such as increase in temperature, increased unpredictability of seasonal rainfall, increase in sea surface temperature, sea level rise, and an increase in extreme events. This will increase the risk of some activities in coastal areas such as fisheries, tourism, and entrepreneurship as well as threaten coastal infrastructure.

At the local level, in particular at villages or communities, the impacts of climate change on the people are site specific. Often they become vulnerable when their livelihood and their places to settle and work are affected by climate disturbances and anomalies. Therefore, involvement of local people to develop local level adaptation plans is necessary. The participatory process starts from the identification of the source of vulnerability, the assessment, development of adaptation planning and action, and finally setting the monitoring and evaluation framework. The vulnerability assessment used a participatory process at the village level. The identification of exposure, sensitivity, and adaptive capacity in the village context was conducted in two tiers by using two matrices. The first matrix plots exposure and sensitivity, which produce potential impacts (Table 4.5). With the framework, the result was plotted to adaptive capacity as illustrated in the second matrix, to produce the level of vulnerability (Table 4.6).

Table 4.5 First Matrix, integration of exposure and sensitivity to show potential impacts

		SENSITIVITY		
		Low	Medium	High
EXPOSURE	Low	Low	Low	Medium
	Medium	Low	Medium	High
	High	Medium	High	High

Table 4.6 Second Matrix, integration of potential impact and adaptive capacity to show vulnerability

		ADAPTIVE CAPACITY		
		Low	Medium	High
POTENTIAL IMPACT	Low	Low	Low	Medium
	Medium	Low	Medium	High
	High	Medium	High	High

Table 4.7 Summary of vulnerability assessments in West Nusa Tenggara Barat and Southeast Sulawesi

	Number of villages studied	Villages with high vulnerability	Villages with medium vulnerability	Villages with low vulnerability
West Nusa Tenggara	50	15	31	4
Southeast Sulawesi	50	9	24	17

The process of assessment actually can be separated into seven steps, which are (i) Developing the village profile (incl. livelihoods, seasonal calendar, map, etc.); (ii) Examining exposure; (iii) Exploring sensitivity; (iv) Understanding village people's adaptive capacity; (v) Assessing vulnerability; (vi) Choosing 1–3 priority problems; and (vii) Planning action plans for solving the 1–3 priority problems. The vulnerability assessments for West Nusa Tenggara and Southeast Sulawesi resulted in the delineation of about 100 villages in both areas into three levels of vulnerability (Table 4.7).

4.3 Adaptation Priorities

4.3.1 Proposed Approach

The global spread and standardization of climate change policies have ignored national vulnerabilities and contexts, where the emphasis is on the same critical

Table 4.8 Summary of differences in vulnerability assessments at multiple levels

National and regional levels	Local (village or community) level
• Hazard risk approach • Comparative study, vulnerable "hot spots" • Conventional socioeconomic data commonly available at statistics office • Unable to understand the causes of social vulnerability • Climate stimuli: climate extremes (hazard related)	• Human capacity (development) and ecosystem service approaches • Comprehensive social or livelihood system data • Climate stimuli: unpredictable weather, climate extremes (livelihood related)

sectors and objectives despite the variability in socioeconomic status, climate risks, vulnerability levels, and capacity (Alves et al. 2020). Similarly, the vulnerabilities identified at the national and regional (sub-national) levels may not link well with vulnerabilities at the local level, in particular at the community level. The approaches and objectives of vulnerability are different at different levels of assessment.

Vulnerability assessments at the national and regional levels in Indonesia commonly use climate-hazard risk approach. This approach successfully identifies vulnerable "hot spots" and is able to compare the degree of vulnerability between regions. However, this approach fails to understand the causes of social vulnerability, in particular the livelihood vulnerability. On the other hand, the human capacity approach works well to understand the cause of social vulnerability at local levels. This means that even though the "hot spots" areas with high vulnerability can be identified from the national and regional approach, a comprehensive vulnerability assessment is necessary to be conducted at the local level in order to better understand the causes of social vulnerability that is site specific and livelihood context dependent (Table 4.8). The importance of local level assessments for addressing complexities that arise from context specificity to determine effective adaptation options is becoming increasingly apparent. Such complexities relate to risk evaluation, downscaled climate projections, impact modelling as well as dynamic projections of vulnerability and exposure, among others (Toimil et al. 2020).

The causes of vulnerability at local levels are often specific; determined by how climate stimuli affect livelihoods of the communities and places where they stay or work. Two approaches are proposed for local level vulnerability assessments and climate change adaptation. These are the Ecosystem-Based Adaptation (EBA) and Community-Based Adaptation (CBA). The EBA focuses adaptation on ecosystem services and their socioeconomic benefits to people or community, through ecosystem conservation and rehabilitation. When coupled with the human factor and systems, this approach offers a comprehensive approach in addressing the climate change issues (Locatelli et al. 2008). The CBA focuses on causes of socioeconomic capacity development as a means of adaptive capacity enhancement through human development. This approach links strongly to people's livelihoods. By this approach the drivers of vulnerability can be emphasized to meet the needs of vulnerable people (Forsyth 2013).

Both EBA and CBA have been conducted in Indonesia on small scales, in various contexts. One example is the study of vulnerability of villages to climate changes at Sentarum Lake in Kalimantan (Santoso et al. 2011). The study is an example of the CBA approach. It shows the dependency of the villagers to regular climate. When excessive rainfall occurs, the villagers look for alternative livelihoods during the extreme excessive rainfall periods and means to process and preserve their agriculture produces. Adhuri et al. (2014) show a vulnerability assessment and adaptation using both the EBA and CBA approaches. The people living in the coastal areas proposed mangrove re-plantation to overcome coastal land erosion from high tide waves (i.e. ecosystem enhancement) as well as a participatory method to build dykes or seawalls, and an introduction of aquaculture technology for community economic development.

Both examples of local and community level adaptations used secondary data of climate change trends, and not the proper climate scenarios. Therefore, these two vulnerability assessments reflect the current vulnerability based on short historical or current climate data. It is arguable that adaptation options that are based on historical and current climate may not be optimal for future climate and lead to lack of adaptation or maladaptation (Toimil et al. 2020). Therefore, climate scenarios are necessary to better select and design adaptation options to better adapt to future climate conditions.

4.3.2 Potential Local Level Pilot

Two areas are proposed to implement enhancement of local level climate change adaptations in Indonesia. The first is in Lombok Island, West Nusa Tenggara. The other is Probolinggo, North Coast of the East Java Province (Fig. 4.5).

Fig. 4.5 Location of proposed projects in Lombok Island and Probolinggo City

Lombok, West Nusa Tenggara is expected to experience serious water problems, a decrease in agriculture production and severe hazards (BAPPENAS 2010a). Its coastal areas are also vulnerable to sea level rise and high tide waves. The area has been identified as highly vulnerable area. Lombok is also classified as a highly vulnerable at the national scale assessment, particularly in the eastern region (Yusuf and Francisco 2009).

The project is to focus on enhancing adaptations of vulnerable local coastal communities in Lombok. The adaptive capacity enhancement requires community or villager participation in the social and ecosystem services diagnosis processes. The project is to be conducted by the Indonesian Institute of Sciences -LIPI (Since Octobre 2021, LIPI merged into National Research and Innovation Agency- BRIN) with its partners. Potential partners are The Ministry of Marine and Fisheries, Ministry of Environment and Forestry, Local Government (province and district levels), academicians (Bogor Agricultural University and Mataram University), local NGOs, local communities or villages, and previous partners in the network. LIPI has some experience working in Lombok with its partners on several research projects in relation to climate change, particularly on the impact of climate change on people migration, food security, and ecosystem approach of adaptations for fisheries management.

Probolinggo on the North Coast of the East Java Province has been identified as a highly vulnerable area with increasing frequencies of climate related hazards (BAPPENAS 2010a). Currently, many parts of the coastal area already undergo coastal abrasions and floods that affect the livelihood of local communities and settlements. Climate change is projected to increase vulnerability of the communities from increasing frequency of floods and expansion of their areas, in particular along the Bengawan Solo and Brantas River banks. Sea level rise and coastal abrasion is expected to become more severe. Irregular and unpredicted seasonal weather is expected to affect the climate-dependent livelihoods such as agriculture, aquaculture, and fishing. The project is to focus on enhancing adaptations of vulnerable local coastal communities in Probolinggo area. The adaptive capacity enhancement requires community or village participation in the social and ecosystem services diagnosis processes. The project is to be conducted by LIPI with their partners. Potential partners are the Ministry of Marine and Fisheries, Ministry of Environment and Forestry, Ministry of Public works, Local Government (province and districts levels), Airlangga University, Local NGOs, Local Communities or villagers, and previous partners in the network. LIPI has some experience working in the Probolinggo area on social vulnerability and climate change projects under National Priority Research Grant no 9 (PN9) on climate change, and already maintains its network (e.g. Adhuri et al. 2014; Wahyono et al. 2013).

4.4 Concluding Remarks

The National Action Plan of Indonesia (NAP/RAN-PI—*Rencana Aksi Nasional Perubahan Iklim*) served as the basis for raising awareness on climate change mitigation and adaptation. This led to strategies and plans for climate action to strengthen institutional capacity. Vulnerability assessments were conducted at the national, regional, and provincial levels with support from multiple donor agencies. Vulnerability assessments at the national and regional levels commonly used the climate-hazard risk approach to identify "hot spots" that enabled regional comparison. However, the approach was found to be limited in revealing the causes of social vulnerability, which is site specific and context dependent. The human capacity approach was better suited for understanding social vulnerability and this was used for local level assessments. Challenges associated with local level assessments relate to risk evaluation, downscaled climate projections, and impact modelling, among others.

Local level vulnerability is often specific and determined by climate impacts on the livelihoods of exposed communities. Two approaches are proposed for local level vulnerability assessments and climate change adaptation. These are the Ecosystem-Based Adaptation (EBA) and Community-Based Adaptation (CBA). The EBA focuses adaptation on ecosystem services and their socioeconomic benefits to people or community, through ecosystem conservation and rehabilitation. The CBA focuses on causes of socioeconomic capacity development as a means of adaptive capacity enhancement through human development. This approach links strongly to people's livelihoods. Two areas are proposed to enhance local level climate change adaptation in Indonesia. These are Lombok Island in West Nusa Tenggara and Probolinggo of the East Java Province.

Acknowledgements The authors acknowledge the ASEAN-India Green Fund for supporting the Project on "Enhancing Local Level Climate Change Adaptation in Southeast Asia" (Phase 1). This chapter draws partly on findings of the Project, specifically the local level needs assessment and scoping exercise conducted in 2015.

References

Adhuri DA, Indarto H, Indrawasih R, Sudiyono, Windarsih A (2014) Membangun dari bawah: strategi adaptasi dan mitigasi perubahan iklim di komunitas pesisir. In: Santoso H. (eds.) Membangun strategi adaptasi dalam menghadapi perubahan iklim. Pusat Penelitian Geoteknologi, Bandung Indonesia, pp 23–139

Alves F, Filho WL, Casaleiro P, Nagy GJ, Diaz H, Al-Amin AQ, Guerra JBSOA, Hurlbert M, Farooq H, Klavins M, Saroar M, Lorencova EK, Suresh J, Soares A, Morgado F, O'Hare P, Wolf F, Azeiteiro UM (2020) Climate change policies and agendas: facing implementation challenges and guiding responses. Environ Sci Policy 104:190–198

BAPPENAS (2008) National development planning: Indonesia response to climate change. Yellow book. BAPPENAS, Jakarta, Indonesia

BAPPENAS (2010a) Indonesia climate change sectoral road map–synthesis report. BAPPENAS, Jakarta, Indonesia

BAPPENAS (2010b) National action plan for greenhouse gas emission reduction (RAN-GRK). Indonesia National Planning Agency, Indonesia

BAPPENAS (2011) Indonesia adaptation strategy: improving capacity to adapt. National Indonesia Planning Agency, Indonesia

BAPPENAS (2012) Strategy to mainstream adaptation into national development plan: integration framework. National Indonesia Planning Agency, Indonesia

DGCC (2015) A glance at directorate general of climate change. Ministry of Environment and Forestry, Jakarta

DNPI (2010) Adaptation science and policy study: book 1 final report. Dewan Nasional Perubahan Iklim: 72

DNPI (2011) Rencana aksi nasional adaptasi perubahan iklim Indonesia. Dewan Nasional Perubahan Iklim, Jakarta Indonesia

Forsyth T (2013) Community-based adaptation: a review of past and future challenges, WIREs Clim Change. https://doi.org/10.1002/wcc.231

Impron BE, Sugiato Y, Muhamad A (2012) Penyusunan basis data dan informasi adaptasi perubahan iklim. Pokja Adaptasi DNPI

KLH (2009) Indonesia second national communication under the United Nations framework convention on climate change (UNFCCC). Ministry of environment. Republic of Indonesia, Jakarta

Locatelli B, Kanninen M, Brockhaus M, Colfer CJP, Murdiyarso D, Santoso H (2008) Facing an uncertain future: how forests and people can adapt to climate change. CIFOR Publications

MoE (2007) Rencana Aksi Nasional dalam menghadapi Perubahan Iklim. Kementrian Lingkungan Hidup, Jakarta, Indonesia

MoE (2010) Indonesia second national communication under the United Nations framework convention on climate change (UNFCC). State minister of environment

MoEF (2018) Indonesia third national communication under the United Nations framework convention on climate change (UNFCC). Ministry of environment and forestry. Jakarta, Indonesia

Santoso H, Tjiu A, Muhammad A (2011) Vulnerability of ecosystem dependent villagers to climate variability: a case study from two villages by the Sentarum Lake, Kalimantan. Workshop on tropical wetland ecosystems of Indonesia: science needs to address climate change adaptation and mitigation. Bali, Indonesia. https://www.forestsclimatechange.org/fileadmin/tropical-workshop/Plenary-4/30A_SantosoH_Vulnerability%20of%20ecosystem.pdf. Accessed 11 December 2012

Toimil A, Losad IJ Nicholls RJ Dalrymple RA Stive MJF (2020) Addressing the challenges of climate change risks and adaptation in coastal areas: a review. Coast Eng 156:103611. https://doi.org/10.1016/j.coastaleng.2019.103611

Wahyono A, Imron M, Nadzir I (2013) Studi penyusunan indikator kerentanan sosial budaya masyarakat pesisir akibat perubahan iklim. Laporan Penelitian Geoteknologi Program Prioritas Nasional No. 9 tahun 2013, Puslit Geoteknologi LIPI

Yusuf AA, Francisco H (2009) Climate change vulnerability mapping for Southeast Asia. Economy and environment program for Southeast Asia (EEPSEA). Singapore

Chapter 5
Climate Change Adaptation in Lao PDR

Vilakone Maniphousay

Abstract Lao PDR is highly vulnerable to the impacts of climate change. In the last three decades, Lao PDR has experienced more intense and frequent floods, and droughts. In the Lao PDR National Strategy on Climate Change 2010, floods and droughts were identified as priority areas for adaptation, which have greatly affected the agriculture, forestry, water resources, and health sectors. Vulnerability assessment for Lao PDR will provide crucial information for effective adaptation strategies. Improving weather forecasting and development of early warning systems would be beneficial to mitigate the impacts of climate change, especially for the vulnerable sectors that have been identified. This will prepare the people of Lao PDR for climate hazards such as floods and droughts, and also reduce the impact of these hazards. The Se Done Basin, which covers the provinces of Saravane and Champasak in the southern part of Lao PDR is proposed as a local level pilot, where the focus is on strengthening the water resource information system.

Keywords Climate change adaptation · Water resources · Vulnerability · Disaster risks · Lao PDR

5.1 Introduction

The Lao People's Democratic Republic (Lao PDR) is a landlocked country in Indochina surrounded by Cambodia, China, Myanmar, Thailand, and Vietnam. In 2012, Lao's administrative hierarchy was divided into 17 provinces and 145 districts and 8,600 villages (Lao PDR 2012). The capital city is Vientiane (Fig. 5.1). The total land area is 236,800 km^2, stretching more than 1,700 km from the north to south. Almost 50% of the country is covered by forest. With no access to the sea, Mekong River, the main river in Laos, shapes a large part of the western boundary with Thailand, while the mountain forms most of the eastern border with Vietnam. Lao PDR is rich in natural resources, namely land and water, forest and mineral resources,

V. Maniphousay (✉)
Department of Climate Change (DCC), Ministry of Natural Resources and Environment, Vientiane, Lao PDR

J. J. Pereira et al. (eds.), *Climate Change Adaptation in Southeast Asia*, Disaster Risk Reduction, https://doi.org/10.1007/978-981-16-6088-7_5

Fig. 5.1 Map of Lao PDR

which are significant potential contributors to national socioeconomic development. The population of the Lao PDR in 2012 is approximately 6.5 million consisting of 49 ethnic groups, and it has a projected growth rate of 2% per annum. Its population density is 28 people/km^2 but the density varies among the provinces. For example, in Phongsaly, Attapeu, and Sekong Provinces the density is about 11–13 people/km^2 respectively, while in Vientiane it increases to 203 people/km^2 (Lao PDR 2012).

Economic growth over the past decade is approximately 6–8% per annum. In the year 2007–2008, the GDP per capita was USD 818, and in 2008–2009 it was USD 906. In 2009–2010, the GDP per capita reached USD 1,069 and GDP per capita estimation for 2015 is approximately USD 1,700. In the Seventh Five-Year Socio-Economic Development Plan, the agriculture and foresty sector are expected to grow at 23% of GDP while the industrial sector is set to grow at 39% of GDP and the services sector is 38% of the GDP compared with 2015 (Lao PDR 2011). Approximately 28% of the population is still in poverty. According to the National Poverty Eradication Programme (NPEP), Lao PDR will eradicate poverty and achieve a millennial goal of exiting the least developed country status by 2020. The Lao PDR's human development index (HDI) has constantly improved over the last decade. In 2014, Lao PDR was in the medium human development category, ranking 139 out of 187 countries and territories in the world. The average HDI for countries in the medium human development group is 0.614 and 0.703 for countries in East Asia and the Pacific region.

5.1.1 Climate

Lao PDR's climate is greatly influenced by tropical typhoons from the southwest and monsoons from the northeast, resulting in heavy rainfall and high humidity. The weather comprises two seasons: wet or rainy season from May to September/October, with a mean annual precipitation of 1,300–3,000 mm, and dry season from late October to April, with an average annual precipitation of 1,950 mm. Temperature generally ranges from 20 °C in the northern and eastern mountainous regions and on the high plateau, to 25–27 °C in the lowlands and the plains.

Rainfall data has been collected consistently since 1980 from four key stations, i.e. Luangprabang, Vientiane, Savannakhet, and Pakse (Table 5.1). In 2004, only 396.7 mm of rainfall was recorded in the Savannakhet station while the maximum precipitation was reported in Pakse at about 2,694.5 mm in 2006. The lowest average precipitation was approximately 1,460 mm in Luangprabang whereas the

Table 5.1 Annual rainfall

Year	Meteorology station (mm)			
	Luangprabang	Vientiane	Savannakhet	Pakse
1980	1,559.5	2,291.4	1,635.6	1,524.5
1985	1,093.4	1,253.5	1,205.1	2,545.2
1990	1,641.6	1,552.1	1,713.5	1,704.0
1995	1,616.9	2,019.8	1,342.3	1,647.5
1996	1,601.0	1,756.0	1,938.0	2,193.0
1997	1,180.0	1,629.6	1,335.2	2,604.4
1998	1,163.7	1,477.4	1,080.4	1,733.5
1999	1,352.6	2,170.7	2,357.1	2,446.6
2000	1,486.7	1,499.8	1,557.8	2,598.4
2001	1,795.0	1,659.0	1,919.9	2,348.6
2002	1,601.8	1,846.7	1,982.0	2,478.0
2003	1,399.0	1,481.0	1,492.3	2,029.1
2004	1,472.7	1,629.6	396.7	1,977.9
2005	1,435.0	1,667.8	1,768.2	1,956.1
2006	1,205.6	1,930.3	1,398.7	2,694.5
2007	1,295.0	1,667.5	1,444.7	1,967.5
2008	1,708.7	2,201.6	1,565.7	1,907.6
2009	1,259.4	1,482.8	1,565.7	2,209.6
2010	1,368.6	1,794.2	1,658.9	1,464.3
2011	2,233.5	2,202.4	1,717.9	2,182.5
2012	1,259.4	1,735.6	1,034.9	1,607.4

Source Lao Statistics Bureau (2012)

highest average precipitation was roughly 2,086 mm in Pakse. The precipitation in Savannakhet fluctuates; it was reported to be about 2,350 mm in 1999 while it was below 500 mm in 2004.

The maximum average temperature at the four stations was above 30 °C whereas the minimum average temperature was below 25 °C (Table 5.2). In Vientiane , Savannakhet, and Pakse stations, the minimum average temperature was above 20 °C, except for Savannakhet station in 1985 where it was below 20 °C. The Luangprabang station was cooler compared with other stations, where it was mostly 20 °C or lower. The trend of mean temperature has shown variations in the Lower Mekong River Basin over the past 60 years, from 1951 to 2010 (Fig. 5.2). In the northern part (Phongsaly, Luangnamtha, and Samnua Provinces) it increased around 0.14–0.23 °C whereas in the central part of Lao PDR it did not change too much, only about 0.04–0.1 °C, and in the southern part, it changed approximately 0.14–0.18 °C.

Table 5.2 Annual average temperature

Year	Meteorology station (°C)							
	Luangprabang		Vientiane		Savannakhet		Pakse	
	Max	Min	Max	Min	Max	Min	Max	Min
1980	31.9	20.5	31.5	22.4	32.0	22.1	32.0	23.2
1985	34.9	16.4	34.2	21.0	35.4	18.0	33.6	20.3
1990	31.2	20.0	31.0	22.6	29.4	24.2	32.1	23.0
1995	31.2	20.0	31.0	22.6	29.4	24.2	32.1	23.0
1996	30.8	20.2	30.9	22.2	31.3	21.8	31.8	23.3
1997	31.6	20.3	31.4	22.6	31.4	21.6	32.1	23.1
1998	32.5	20.2	32.5	23.2	32.7	22.8	33.1	24.4
1999	30.8	20.2	30.9	22.2	31.3	21.8	31.8	23.3
2000	31.4	20.1	31.4	22.2	30.3	20.8	32.3	23.4
2001	31.6	20.9	31.7	22.6	31.4	21.5	32.8	23.1
2002	30.9	20.3	31.6	22.5	32.4	22.0	33.1	23.6
2003	32.0	19.9	31.8	22.3	32.4	22.6	32.5	22.8
2004	31.7	19.4	31.2	22.1	31.7	21.1	32.4	23.1
2005	32.2	19.8	31.8	22.7	31.8	22.0	32.4	23.7
2006	32.6	20.3	31.7	22.5	32.2	22.1	32.5	23.5
2007	31.2	20.0	31.0	21.9	31.3	21.1	31.8	22.7
2008	31.1	20.1	30.5	22.0	30.0	20.2	31.8	22.5
2009	32.6	20.2	31.9	22.4	30.3	20.9	32.4	23.2
2010	33.4	20.7	32.2	22.9	32.3	21.3	33.1	23.9
2011	31.4	20.1	30.8	22.4	30.7	20.6	31.9	22.6
2012	32.8	21.1	32.0	23.7	32.0	21.9	33.0	23.6

Source Lao Statistics Bureau (2012)

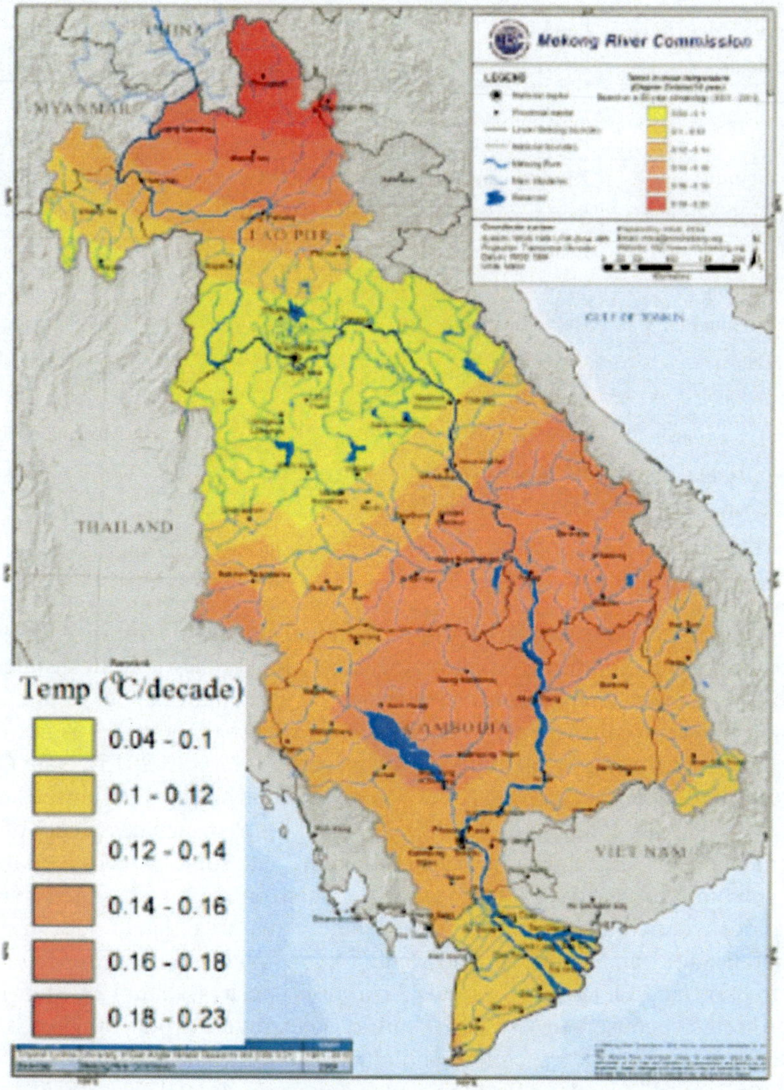

Fig. 5.2 Change of temperature in the Lower Mekong River Basin, 1951–2010. *Source* MRC (2010)

5.1.2 Natural Resources

Lao PDR has plenty of surface water and the Mekong River is the main waterway, flowing from the north to the south of the country (Table 5.3). Water from many small rivers and tributaries represents a vital natural resource for national socioeconomic development, particularly for agriculture and hydroelectric generation. Agriculture is the most important economic sector for the country, while the generation of electricity

Table 5.3 Main rivers in Lao PDR

River name	Flowing by	Length (km)	Remark
Mekong	Laos	1898	Main river
Nam Ou	Phong saly-Luangprabang	448	Mekong tributary
Nam Ngum	Xiengkhuang-Vientiane	354	Mekong tributary
Nam XeBangHieng	Savannakhet	338	Mekong tributary
Nam Tha	Luangnamtha-Bokeo	325	Mekong tributary
Nam Xekong	Saravane-Sekong-Attapeu	320	Mekong tributary
Nam XeBangPhay	Kahmmuane-Savannakhet	239	Mekong tributary
Nam Beng	Ourdomxay	215	Mekong tributary
Nam XeDone	Saravane-Champasack	192	Mekong tributary
Nam XeLaNong	Savannakhet	115	Mekong tributary
Nam KaDing	Borikhamxay	103	Mekong tributary
Nam Khane	Huaphang-Luangprabang	90	Mekong tributary

Source Lao Statistics Bureau (2012)

from hydropower serves as the core of the national strategy on turning Lao PDR into the battery of the Southeast Asian region. The annual renewable freshwater is 270 billion cubic metre and water usage is shared by agriculture (about 82%), followed by industry (approximately 10%) and domestic supply (about 8%). Water usage in the other sectors is negligible (http://www.wepa-db.net, April 2015). Nam Ou is the longest tributary of the Mekong River whereas Nam Khane is the shortest tributary; both are located in Luangprabang. The Savannakhet Province has plenty of surface water from Nam XeBangHieng, Nam XeLaNong, and some parts of XeBangPhay.

According to the Land Law (Art.11), land is to be classified into eight categories: agricultural land, forest land, water-area land, industrial land, communication land, cultural land, land for national and security defence, and construction land. Of Lao PDR's total 23.6 million hectares, about 17 million hectares are protected or conserved forests, 1.8 million hectares are production forest, and the remaining areas are for agriculture and other land uses. Nearly 70% of the agricultural land is paddy land, while the balance comprises annual crops, fruit trees, and grassland (Fig. 5.3). Maize is increasingly replacing upland rice, and areas planted with maize nearly tripled between 2005 and 2010, from 86,000 hectares to more than 212,000 hectares. Such expansion results in substitution of other crops or forest encroachment. Because maize is planted in the same plot every year, this is exposing the land to high soil erosion and unsustainable use.

Forests are an important natural resource in Lao PDR, providing timber and non-timber products. Furthermore, it provides a range of environmental services, for instance, waterway protection, earth surface preservation, biodiversity habitats, and other ecosystem services. It is recognized that forests play an important role not only in climate change mitigation but in climate change adaptation. Forests contribute significantly to socioeconomic development in Lao PDR, especially in the rural area

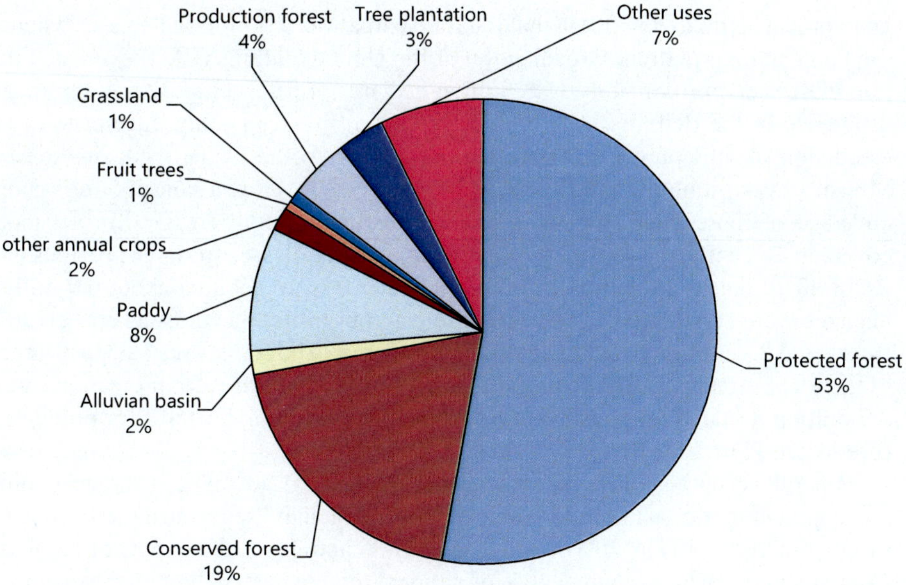

Fig. 5.3 Distribution of land resources by type in 2009

where most poor people live and depend heavily on the forest for food, fuel, and shelter. Although forests are important for the people's livelihoods, deforestation and forest degradation have been accelerating rapidly in the last few decades. About 41.5% of the country's land area was covered by forest in 2002, compared with 64% in 1960 and 70% in 1943 (Fig. 5.4) (Ministry of Agriculture and Forestry 2005). The Lao

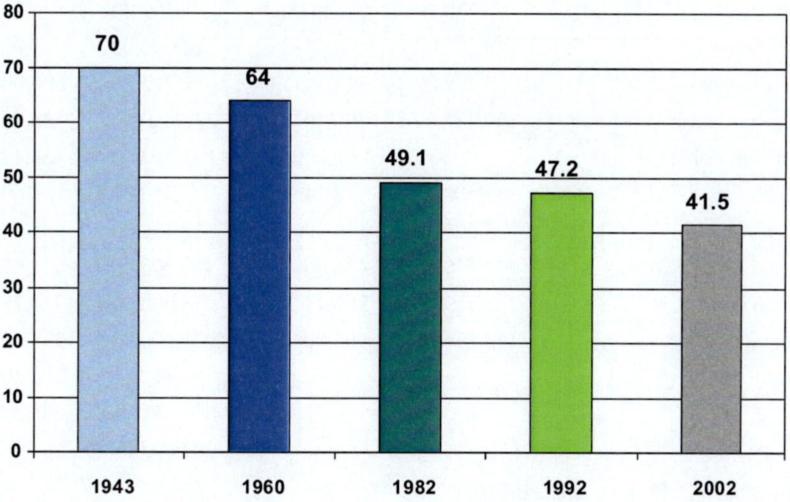

Fig. 5.4 Percentage of forest cover in Lao PDR. *Source* Ministry of Agriculture and Forestry (2005)

government estimated in 2010, that the forest area was 9.5 million hectares whereas the Food and Agriculture Organization of the United Nations (FAO) estimated that Lao PDR had approximately 15.8 million hectares of forests in 2010. A significant difference in the definition of forest cover has resulted in a big difference in the estimation of the country's forest area. According to the Department of Forestry, Ministry of Agriculture and Forestry, 2005 Lao PDR, "forest" means tree canopy coverage of more than 20% while the international definition is 10% of canopy coverage. Recently, the forest was classified into three classes from the previous five. As of 2010, out of a total of 13.5 million hectares of forest area about 6.9 million hectares were considered as protection forest, 3.6 million hectares as conservation forest, and 3.1 million hectares as production forest. The forest area category targets of the forest strategy 2020 are proposed as 8.2 million hectares for protection forest, 4.7 million hectares for conservation forest, and 3.1 million hectares for production forest (Lao PDR 2005).

Mineral resources have become significant for Lao PDR's socioeconomic development in the last decade (Table 5.4). Mining activities contributed about 1% to GDP in 2000 (STEA 2000), whereas in 2006 and 2007, the export of Gold and Copper had reached more than 40% of national total exports. The mining sector is a major contributor to GDP growth. Lao currently approved 178 mining projects by a total of 127 companies, 85 of which are foreign owned (Department of Mines 2008).

Table 5.4 Minerals extracted in Lao PDR, 2005–2008

Mineral	2005	2006	2007	2008
Gold + Silver (tonnes)	10.15	12.65	9.09	5.81
Copper (tonnes)	30,480.00	60,804.00	62,452.00	64,705.00
Copper concentrate (tonnes)		746.00	3,358.81	107,430.48
Tin (tonnes)	240.00	809.00	1,109.00	260.00
Barite (tonnes)	8,600.00	6,500.00	10,000.00	1,000.00
Lignite (tonnes)	358,637.88	319,242.00	681,715.00	600,000.00
Gypsum (tonnes)	109,787.00	206,104.00	232,250.00	586,866.00
Anthracite (tonnes)		62,000.00	131,972.00	158,490.17
Lead (tonnes)	3,500.00	2,249.00	2,350.00	600.00
Zinc (tonnes)	8,000.00	4,000.00	12,116.00	6,079.00
Limestone (tonnes)	190,440.00	430,000.00	450,000.00	911,654.00
Gravy (cu.m.)	58,719.00	900,150.00	943,000.00	214,772.00
Sand (cu.m.)	160,136.00	900,000.00	920,000.00	208,608.00
Construction M (cu.m.)				70,210.00
Soil (cu.m.)				4,012.00
FE (tonnes)			20,000.00	36,218.00

Source Ministry of Energy and Mines (2008)

5.2 National Perspective

5.2.1 National Arrangements and Institutions

Lao PDR ratified the United Nations Framework Convention on Climate Change (UNFCCC) in 1995 and the Kyoto Protocol in 2003. The climate change issue has been mainstreamed into the 7th National Socio-Economic Development Plan (NSEDP-7) as well as other strategies, for example, the national environment, forest, agriculture, energy strategies, and other sectors. In 2008, the National Steering Committee on Climate Change (NSCCC), chaired by the Deputy Prime Minister, was established to formulate climate change strategies, programmes, and projects for Lao PDR. In 2013, Joint Crediting Mechanism Committee and Technical Working Group on Climate Change were established with eight cross-sectoral working groups. The Ministry of Natural Resources and Environment (MoNRE) has been designated as the national focal point for climate change matters. The national strategy on climate change was approved in early 2010 with a vison "to secure a future where Laos is capable of mitigating and adapting to climatic condition change in a way that it promotes a sustainable economic development, reduces poverty, protects public health and safety, enhances the quality of Lao PDR's natural environment, and improves the quality of life for all Laotians".

Lao PDR has faced significant threats from climate change in recent years. More than 70% of its population are highly dependent on natural resources for their livelihood and they will be exposed to climate risk in the future. Floods and droughts have greatly impacted on agriculture, forestry, water resources, and health. These sectors were identified as priority areas for adaptation in the national strategy on climate change 2010. Key environmental legislation and social considerations give emphasis to sustainable development (Table 5.5).

The Climate Change Office, which is under the Department of Environment (Water Resources and Environment Administration, WREA) was established in 2008. The Office is responsible for climate change issues in the country. In November 2011,

Table 5.5 The Major existing laws and regulations

Title	Enacted year	Code
Environmental protection law	1999	No.02/NA
Forestry law	2007	No.06/NA
Regulation on environmental assessment in the Lao PDR	2000	No. 1770/STEA
Decision on the environmental impact Assessment for industries and handicrafts Processing in the Lao PDR	2005	No.1222/IH
Decree on the environmental impact assessment in the Lao PDR	2010	No.112/PMO.WREA

the Department of Disaster Management and Climate Change (DDMCC) was established under the Ministry of Natural Resources and Environment (MoNRE). The DDMCC has been assigned to study and enhance, disseminate and implement policies and strategies related to disaster and climate change matters (Fig. 5.5). Another mandate of DDMCC is to coordinate with line ministries and all related stakeholders at both national and local (provincial) levels as well as act as the focal point for all disaster and climate change matters (DDMCC 2013a). The Technical Working Group on Climate Change (TWGCC) was established on February 2013. This working group consists of DDMCC as a leader and representatives of Ministry of Agriculture and Forestry, Ministry of Energy and Mines, Ministry of Science and Technology, Ministry of Public Works and Transport, Ministry of Industry and Commerce, Ministry of Health and Ministry Education, and Sport as committee members. TWGCC will act as the secretariat for MoNRE and the National Environment Committee (NEC). TWGCC will also study and provide comments on policies, strategies, plans, programmes, and other issues related to climate change, and will be the focal point for those sectors at national and provincial level in relation to climate change issues.

In November 2013, the Secretary of the National Disaster Management Committee, National Disaster Management Office, Ministry of Labour, and Social Welfare was transferred to the DDMCC (MoNRE). The NEC was established by Prime Ministerial Decree to direct environmental management and address environmental issues in a uniform manner. Provincial authorities established their own provincial environmental committees (PECs). Lao PDR has streamlined the coordination bodies for disaster prevention and control (Fig. 5.6). The Deputy Prime Minister is the chair of National Disaster Prevention and Control Committee and the DDMCC serves as the secretariat to this Committee. The Committee has revised its mandate three times. Most recently, sub-national level Disaster Prevention and Control Committee have been established for 14 provinces, 26 Districts, and 225 Villages (Kaisorn 2015).

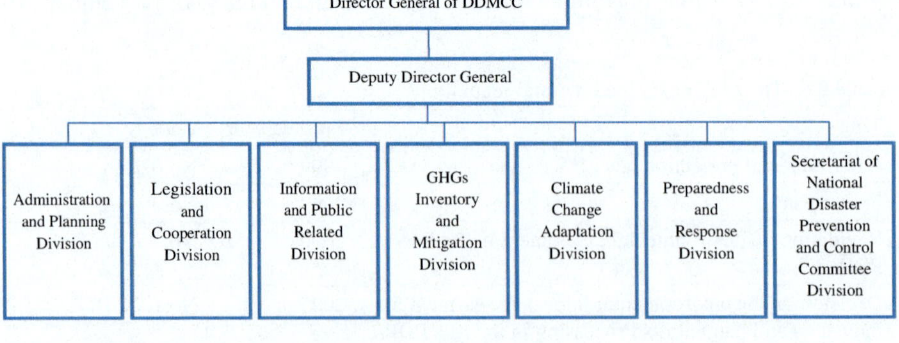

Fig. 5.5 Key divisions in the department of disaster management and climate change

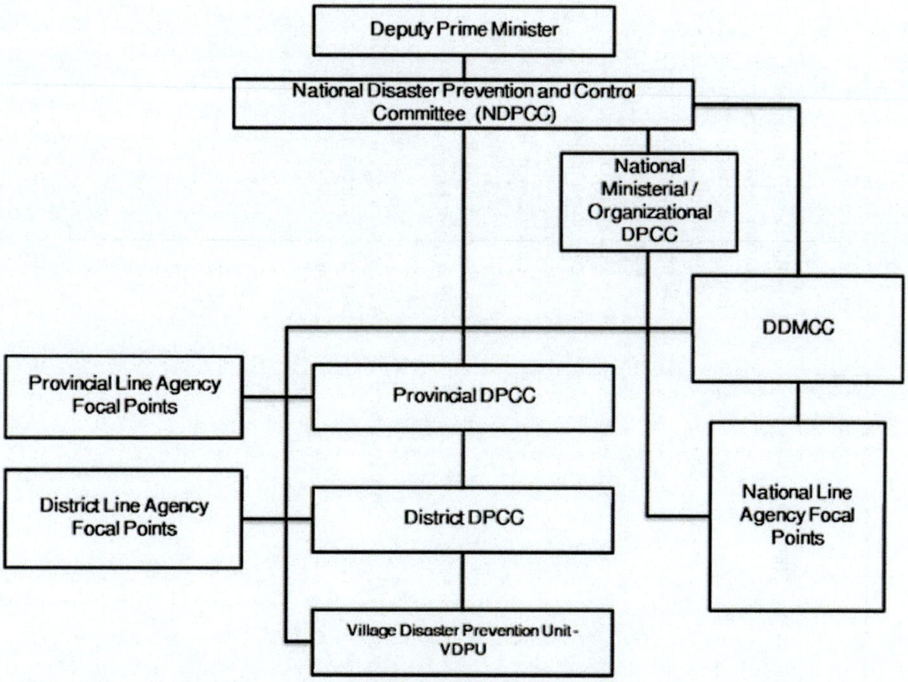

Fig. 5.6 Coordination structure for disaster management in Lao PDR

The DDMCC fully supports the inclusion of disaster and climate change issues to secondary and university syllabus as there is a pilot project called "ProCEEd—Promotion of Climate-related Environmental Education". This project "aims at improving the knowledge, attitudes, and practices related to the environment , biodiversity and climate change in selected target areas in Laos through communication and education" (http://www.laos-proceed.com). As Lao PDR is a creditor on GHGs emission, Lao PDR supports enterprises and the private sector for involvement in climate change reduction, for example in carbon trading, through the Clean Development Mechanism and Joint Crediting Mechanism (DDMCC 2013b).

Lao PDR has several strategies and documents on climate change (Table 5.6). The First National Communication on Climate Change (FNC) to the UNFCCC was submitted in 2000, and the Second National Communication on Climate Change (SNC) was approved in 2013. The National Adaptation Programme of Action (NAPA) was approved and submitted to the Lao government in 2009; another strategy related to climate change is the National Strategy on Climate Change of the Lao PDR (NSCC) approved by the Lao government in 2010; and the Guidelines on Development and Consideration of Proposed Clean Development Mechanism project in Lao PDR was approved in 2012. In 2013, Climate Change Action Plan of Lao PDR for 2013–2020 was approved by MoNRE; and the Guidelines on Ecosystem-Based Adaptation Practices in Lao PDR were approved by MoNRE in 2014. Key aspects of the documents are as follows:

Table 5.6 Key policies on climate change in Lao PDR

Item	Name of policy action	Government division responsible	Status	Sector(s) of focus	Summary description
1	National communication to the UNFCCC	Department of environment	Submitted November 2000	Multi-sectoral	Provides details on national circumstances with respect to climate change, an inventory of greenhouse gas emissions, as well as identification of impacts of climate change
2	National adaptation programme of action	National environment committee	Submitted May 2009	Multi-sectoral	Provides an update and background on national circumstances and vulnerability to climate change. Identifies key adaptation needs and criteria for priority activities. Concludes with a list of proposals for actions
3	The national strategy on climate change (NSCC)	Water resource and environment agency	Approved March 2010	Multi-sectoral	A framework document which lists priority areas for adaptation and mitigation

(continued)

Table 5.6 (continued)

Item	Name of policy action	Government division responsible	Status	Sector(s) of focus	Summary description
4	The guidelines on development and consideration of proposed clean development mechanism project in Lao PDR	Ministry of natural resources and environment (MoNRE)	Approved November 2012	Multi-sectoral	
5	The second national communication (SNC)	Ministry of natural resources and environment (MoNRE)	Submitted March 2013	Multi-sectoral	Revises national circumstances on climate change, an inventory of greenhouse gas emissions, as well as identification of impacts of climate change and climate change vulnerability assessment, potential options for adaptation actions
6	Climate change action plan of Lao PDR for 2013–2020 (CCAP)	Ministry of natural resources and environment (MoNRE)	Approved April 2013	Multi-sectoral	
7	The guidelines on ecosystem-based adaptation practices in Lao PDR	Ministry of natural resources and environment (MoNRE)	Aproved 2013	Few sectoral	

(continued)

Table 5.6 (continued)

Item	Name of policy action	Government division responsible	Status	Sector(s) of focus	Summary description
8	Seventh national socio-economic development plan (2011–2015)	Ministry of planning and investment	Completed 2010	Multi-sectoral	An overarching planning document for the country. It lists achievements during the past five years and lays the framework for the country's goal of, "national development, to achieve economic growth of at least 8% annually, reduce poverty, achieve the Millennium Development Goals by 2015 and construct basic infrastructure for industrialization and modernization in the times to come." Mitigating the effects of climate change is mentioned as a macro level goal under environmental protection and natural resource management
9	Intended nationally determined contribution, INDC	Ministry of natural resources and environment (MoNRE)	Approved 2015	Multi-sectoral	Provides details on national circumstances with respect to climate change (mitigation and adaptation), identifies what key sectors intend to do in mitigation and adaptation till 2030

- The FNC had not provided climatic vulnerability information but it had greenhouse gas inventory and other mitigation aspects. The adaptation aspects were proposed as a priority activity in the next national communication. The NAPA was submitted to UNFCCC in 2009 by the Ministry of Natural Resources and Environment , MoNRE, (formerly Water Resources and Environment Administration, WREA). The NAPA identified 45 priority projects on climate change adaptation which covered four sectors: agriculture, forestry, water resources, and public health. Twelve project proposals were classified as first priority whereas the rest were considered as second priority projects. Of the total number of 45 projects, 13 proposed projects were from the agriculture sector, 14 were from the forestry sector, 10 were from the water resources sector and 8 were related to public health.
- The NSCC was endorsed by the Prime Minister in 2010. Its main goals are to reinforce sustainable development goals of the Lao PDR. The NSCC has provided both adaptation and mitigation options for seven sectors namely agriculture and food security, forestry and land-use, water resources, energy and transport, industry, urban development, and public health. Mitigation options were not considered for the water resources and public health sectors.
- The Guidelines on Development and Consideration of Proposed Clean Development Mechanism Project in Lao PDR were approved by MoNRE in 2012. It provides principles and stages for development projects for instant hydropower, energy efficiency, bio-gas, and other green development projects that are suitable for Clean Development Mechanism (CDM) criteria for carbon trade or carbon credit. It is under mitigation measures.
- The SNC was submitted to UNFCCC on March 2013 by MoNRE; it reported the climate change situation and provided more climate change information than the FNC. The General Circulation Models (GCMs) were used to simulate scenarios of seasonal rainfall changes and provide vulnerability and risk information for the country. More details on GHGs inventory assessment were provided with a listing of 10 CDM projects (approximately 1,450 Gg of CO^2 potential reduction). The SNC also reported the constraints, gaps, and needs to support the implementation of climate change activities in the country.
- The Climate Change Action Plan of Lao PDR for 2013–2020 (CCAP) was approved by MoNRE in 2013. It was based on the Strategy on Climate Change of Lao PDR and the Second National Communication on Climate Change. The main purposes were to propose projects and activities as well as leading responsible agencies for implementation in the year 2013–2020. There were four key initiatives mentioned: instant strengthening of institutional and human resource capacities on climate change , climate change adaptation, climate change mitigation through the reduction of greenhouse gas emissions, and strengthening education and raising public awareness on climate change.
- The Guidelines on Ecosystem-Based Adaptation Practices in Lao PDR provide a suitable template for analysis of vulnerability and the selection of adaptation

options at the sub-national level in Lao PDR. Climate change policy for the sub-national level is still being considered because the line ministry and climate change agency of MoNRE reaches the provincial level, and climate change issues are still new for them. According to the Environmental Action Plan of Provincial for 2016–2020, each Provincial of Natural Resources and Environment has already mainstreamed climate change issues into their action plans.

5.2.2 Future Plans and Challenges

The Natural Resources and Environment Vision 2030 covers aspects of climate change adaptation and disaster risk reduction, and specifically mentions developing the following instruments: a law on disaster prevention and control and climate change, master plan of climate change adaptation, regulation on disaster management, and guidelines on disaster management in the village, district, and provincial levels (Lao PDR 2015). The Action Plan 2020 of MoNRE has also articulated the following:

- Complete the law on Disaster Prevention and Control and Climate Change; the national strategy and action plan on Disaster Risk Reduction (2016–2020); National Adaptation Plans and National Implementation GHGs Mitigation Plan.
- Complete establishment of Disaster Prevention and Control and Climate Change Committees in both national and sub-national levels.
- Complete guidelines on mainstreaming disaster management and climate change into the 8th socioeconomic development plan (2016–2020) of Lao PDR as well as other strategies and action plans.
- Complete development of climate change adaptation and mitigation mechanisms and implementation of pilot projects on climate change adaptation in agriculture and forestry.
- Complete establishment of a national disaster fund and climate change research fund.

The DDMCC is taking the lead on developing the law on Disaster Prevention and Control and Climate Change. Preparation for the Third National Communication is also underway. The National Adaptation Plan, NAP, which is under development is expected to encompass disaster risk reduction plans for eight provinces, 45 districts, and 160 villages.

Lao PDR is an agricultural country and it depends on natural resources. As such Lao PDR has been recognized as one of the countries that are most vulnerable to climate change impact. Limited knowledge and skills on climate change is a main challenge for Lao PDR. Climate change is a rather new subject to the Lao community and requires technical expertise to develop climate scenarios and impact models in relevant sectors. Capacity has to be strengthened to address gaps in climate and statistical modelling, vulnerability assessment, and adaptation.

The lack of long-term historical data on meteorology, hydrology, water flow, and forest resources constrains vulnerability adaptation as well as in developing good scenarios for climate modelling. Lao PDR experienced a major flood in 2005, which cost approximately USD 28.5 million (Lao PDR 2013). To prevent or mitigate such hazards, improvement of the weather forecast system and development of earlier warning systems should be addressed. This is not only to prepare the Lao people for climate hazards such as floods and droughts, but also to reduce the impact from these hazards.

5.2.3 *Vulnerable Sectors and Regions*

The climate of Lao PDR is dominated by monsoons with high rainfall and high temperatures. The precipitation is approximately 70% during the wet season and the climate is driven by interannual rainfall variability. Extreme climate events may occur with more frequency and more intensity and these extreme events are likely to lead to an increase in climate-related hazards (Lao PDR 2013). It has been suggested that greenhouse forces will result in "more frequent El Nino-like conditions and stronger cold events" (Timmermann et al. 2009). Observation data shows that the top 10 El-Nino events of this century occurred mostly after the 1970s. It appears that extreme climate events in the region and climate anomalies associated with ENSO are being exacerbated by increasingly higher temperatures.

Disasters: Climate-driven disasters that normally strike Lao PDR are floods, droughts, and storms. Flooding is the most frequent hazard resulting in severe economic losses. Between 1966 and 2008, floods occurred 33 times, and the average cost of each disaster was approximately USD 14 million (Table 5.7). Flooding is normally followed by health epidemics such as cholera, diarrhoea, and dengue. Most storms are followed by floods or severe floods while delays in rainfall can lead to droughts. In the last three decades, floods and droughts occurred more intensely and frequently. These events sometimes happened twice in a year. They have significantly impacted the agriculture, forestry, water resources, health, transportation, and energy sectors. The disasters have not only caused serious economic losses, but they have also impacted livelihoods and loss of lives in many cases. The number of people affected fluctuates considerably between both flood and drought events. It is noted that over the years, the number of affected people and economic losses have increased (OFDA/CRED International Disaster Database). In developing countries, many cases of severe climate-related disasters increased the budget deficit by between 0.23 and 1.1 percent of the GDP (World Bank 2010). In Lao, the biggest flood disaster that occurred in the last 30 years is due to the Ketsana tropical cyclone that hit in September 2009. This disaster cost over USD 50 million, which is equivalent to about 1.1% of the Lao GDP. The recovery cost is estimated to be over USD 60 million (Table 5.8).

Drought conditions have been reported in Lao PDR since the 1980s. Since 1966, a total of nine drought disasters have been recorded; the cost of damage has increased

Table 5.7 Damages caused by floods in Lao PDR from 1966 to 2008

Item	Year	Type of damage	Damage cost (USD,000)	Area/region
1	1966	Large flood	13,800	Central
2	1968	Flood	2,830	Central and Southern
3	1969	Flood	1,020	Southern
4	1970	Flood	30	Central
5	1971	Large flood	3,573	Central
6	1972	Flood and drought	40	Central
7	1973	Flood	3.7	Central
8	1974	Flood	180	
9	1976	Flash flood	9,000	Central
10	1978	Large flood	5,700	Central and Southern
11	1979	Flood and drought	3,600	Northern and Southern
12	1980	Flood	3,000	Central
13	1981	Flood	682	Central
14	1984	Flood	3,430	Central and Southern
15	1985	Large flood	1,000	Northern
16	1986	Flood and drought	2,000	Central and Southern
17	1990	Flood	100	Central
18	1991	Flood and drought	3,650	Central
19	1992	Flood, drought and forest fire	302,151.20	Central (F) and Northern (D)
20	1993	Flood and drought	21,827.93	Central and Southern
21	1994	Flood	21,150	Central and Southern
22	1995	Flood	15,000	Central
23	1996	Large flood and drought	10,500	Central
24	1997	Flood and drought	1,860.30	Southern
25	1999	Flood	7,450	Central
26	2000	Flood	6,684.23	Central and Southern
27	2001	Flash flood	808.5	Central and Southern
28	2002	Large flood, flash flood and landslide	14,170	Northern, Central and Southern
29	2004	Flood	750.399	Southern
30	2005	Flash flood and landslide	1,316.58	Central and Southern
31	2006	Flood	3,636	Central and Southern
32	2007	Flash flood	8,056	Northern, Central and Southern
33	2008	Large flood and flash flood	4,384.40	Northern and Central

Source ADPC (2010)

Table 5.8 Costs associated with the Ketsana tropical cyclone of 2009

No	Affected sector	Economic Loss (USD)	Recovery cost (USD)
1	Social sector	10,945,014	22,534,600
2	Productive sector	22,211,376	15,000,000
3	Infrastructure	21,266,000	24,513,750
	Total	**54,422,390**	**62,048,350**

Source Rizaldi et al. (2013)

Table 5.9 Damages caused by drought in Lao PDR from 1966 to 2008

item	Year	Type of damage	Damage cost (USD,000)	Area or country region
1	1967	Drought	5,120	Central and Southern
2	1975	Drought	N/A	Central
3	1982	Drought	N/A	N/A
4	1983	Drought	N/A	N/A
5	1987	Drought	5,000	Central and Southern
6	1988	Drought	40,000	Southern
7	1989	Drought	20,000	Southern
8	1998	Drought	5,763	Northern and Southern
9	2003	Drought	16,500	Central and Southern

Source ADPC (2010)

in the last decade (Table 5.9). Droughts that occurred in 1988 and 1989 were very severe, causing damages up to USD 40 million and USD 20 million, respectively. People who lived in the rural areas were most affected due to drought-reduced crop production as well as water supply. They are mostly farmers or agriculturalists, and their livelihoods have been reliant on natural conditions such as precipitation for cropping and aquaculture.

Vulnerable Sectors: There are four key policies important for climate change adaptation in Lao PDR (Table 5.10). These are the National Adaptation Programme of Action to Climate Change (NAPA), Strategy on Climate Change of the Lao PDR (SCC), Climate Change Action Plan of Lao PDR for 2013–2020 (CCAP), and Technology Needs Assessment Report on Climate Change Adaptation (TNACC). Each of the policy mentions sectors of importance. Four sectors were mentioned in the NAPA, 10 sectors were identified in the SCC, nine sectors were recognized in CCAP, and two sectors were prioritized in TNACC. The agriculture and water resource sectors are mentioned in all four policies; indicating the key vulnerable sectors for Lao PDR. Both agriculture (rice farming, fisheries, and vegetation) and water resources (water supply to domestic consumption as well as cropping, and water for electricity generation) are severely affected by floods and droughts. Other affected sectors include land use and forestry, human health, and infrastructure (irrigations, housing,

Table 5.10 Vulnerable sectors highlighted in key policies of Lao PDR

Item	Sectors	Policies			
		NAPA	SCC	CCAP	TNACC
1	Agriculture	x	x	x	x
2	Forestry	x	x	x	
3	Water resources	x	x	x	x
4	Health (human health)	x	x	x	
5	Food security		x		
6	Land use change		x	x	
7	Energy		x	x	
8	Transportation		x	x	
9	Industry		x	x	
10	Urban development		x	x	

roads, and bridges). The sectors are strongly connected; when one sector is affected by climate threats, the other sectors are also affected. Lao PDR relies heavily on agriculture and agricultural activities, which are dependent on natural factors like temperature, precipitation, and water. Rice farming and livestock rearing are major activities that consume water heavily for good production. Thus, water is a crucial element for higher yields especially in rice and vegetable farming.

The agriculture sector dominates the economic activity of Lao PDR, contributing more than 30% to the country's GDP in 2011. Rice farming, livestock rearing , fisheries, and other agriculture activities are mostly in remote areas or suburbs where basic facilities or infrastructure are limited. These limitations then affect the efficiency of agricultural production. Most cases of floods are associated with storms or tropical cyclones. For instance, the Ketsana tropical cyclone, which brought heavy rains as well as strong winds hit Lao PDR in 2009, and caused severe damage to rice crops and infrastructure such as roads and irrigation (Table 5.8). Agriculture is practised mostly in the suburbs or rural areas where most Lao people live. Lao farmers generally use traditional methods, and they are not well educated. As a consequence, when floods and droughts occur, they cannot cope and most of the productivity of agriculture is widely lost. Such events contribute to food insecurity and put Lao farmers below poverty baseline. The poverty level in some districts or villages worsens after a disaster event. In 2015, there was a locust invasion in the north affecting Phonthong District, Luangprabang Province (Fig. 5.7). The event was believed to be linked to a change in the climate (MAF 2015).

Water resources are widely used in every sector, for instance, agriculture, energy (hydropower), water supply (domestic consumption, factory, or manufacturing). Even if people have surpassed annual water resources usage, they still have to try very hard to manage it and make the most efficient plan for every sector. Water quantity and quality are being reduced by climate hazards and as mentioned in the IPCC report (AR4 2007), the climate hazards events are going to be more intensified

Fig. 5.7 A locust attack was reported in Luangprabang Province in 2015

and frequent. Heavy rains cause floods which increase massive volumes of water, as well as cause soil erosion, and reduce the quality of water. As there are many agencies related to water resources, for instance, the Department of Water Resources, Department of Meteorology and Hydrology, Lao National Mekong Committee Secretariat (MoNRE), Department of Housing and Urban Planning, Water Supply State enterprise (MPWT), Department of Irrigation (MAF), National Centre of Environmental Health and Water Supply (MoH), Department of Energy Policy and Planning (MEM), water resources are cross-cutting sectors. As a consequence, whenever water resources change they will affect many other sectors.

Vulnerable Regions: Vulnerability assessments are important to provide crucial information for developing an adaptation strategy. The assessment contributes to enhance understanding of an area and its constituents such as people, villages, natural resources, agriculture, and forest that should be focused on to address the impacts of climate change. Vulnerability assessments also reveal the existing level of adaptive capacity in sectors or systems, potential impacts of climate change as well as exposed elements, so that appropriate adaptation strategies and interventions can be implemented to enhance resilience.

Vulnerability assessment at the regional level revealed that Lao PDR is one of the countries in Southeast Asia that is vulnerable to the impacts of climate change (Fig. 5.8). In this assessment, most of the northern Bolikhamxay and Savannakhet Provinces were classified as the most vulnerable while other provinces were classified as vulnerable provinces. Vulnerability assessment at the local level indicated that the most vulnerable districts are in the southeastern provinces, which are next to the Vietnam border (Rizaldi et al. 2013). Some districts in the northern part of Lao PDR, i.e. Phongsaly, Luangnamtha, Sayabouly, Oudomxay, LuangPrabang, Samnuea, and

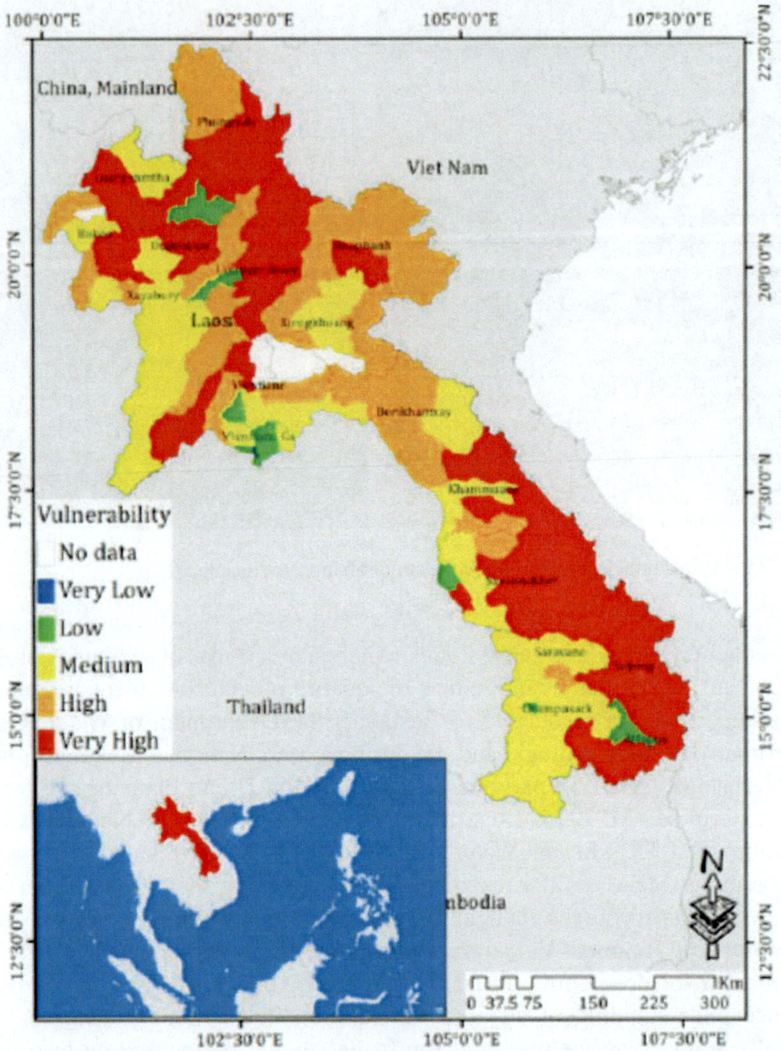

Fig. 5.8 Vulnerability level of districts in Lao PDR. *Source* Rizaldi et al. (2013)

Vientiane Provinces were also classified as most vulnerable. Four inner districts of Vientiane, Pakse (Champasak Province), LuangPrabang districts (LuangPrabang Province), and Thakhek district (Khammouan Province) were classified as having low level of vulnerability to climate change. The designation of LuangPrabang town as a UNESCO World Heritage site warrants further investigation on its vulnerability due to climate change, which is compounded by the challenges of rapid development and environmental conservation (Fumagalli 2020).

5.3 Adaptation Priorities

5.3.1 Approach

The assessment of vulnerable sectors was done by reviewing key climate change policies and conducting interviews with members of the technical working groups on climate change as well as academia. These included officials from the Ministry of Energy and Mines (MoEM), Ministry of Agriculture and Forestry (MAF), Ministry of Health (MoH), Ministry of Education and Sport (MoES), Ministry of Industry and Commerce (MoIC), Ministry of Public Works and Transport (MPWT), as well as academics from the forestry and natural science departments of the National University of Laos (NUoL). The interviews were followed by a consultation workshop covering a wider range of participants, including non-government organizations.

The process resulted in identifying water as a priority sector followed by agriculture. It was agreed that water resources should be emphasized when developing local level adaptation plans. Water resources are a fundamental resource for life and socioeconomic development. Both the quantity and quality of water resources are expected to be impacted by climate change. Therefore, the water resources sector requires enhanced adaptive capacity to cope with the impacts of climate change and be resilient.

5.3.2 Potential Local Level Pilot

The proposed local level pilot is the Se Done Basin, located in the southern part of Lao PDR covering the Saravane and Champasak Provinces (Fig. 5.9). The area is approximately 7,200 km^2. The focus is on "Strengthening Water Resource Information System for Climate Change". Flooding commonly occurs in the Se Done Basin and it causes great damage on rice fields, fish ponds, irrigation, roads, and houses. As agriculture and water resource sectors are recognized as the most vulnerable sectors, this pilot project will support both sectors with water resource information.

Farmers will be informed about projected rainfall and quality of water before they plant their crops. With this assistance, the management of water for various sectors is expected to improve. Furthermore, early warning systems will offer information to people for better response to floods. Rice farming, cassava, coffee, and maize as well as livestock will face increased temperatures, which will reduce crop yields. Increased precipitation will cause flash floods. Such conditions will put pressure on food security, economic development, and livelihood of farmers in the area.

Information management would be an adaptation option in the area to cope with such threats (storms, flash floods, and flooding). Providing better climate information to people in the Se Done basin would enable them to be better prepared to reduce potential impacts. Especially during floods, a better information-management system

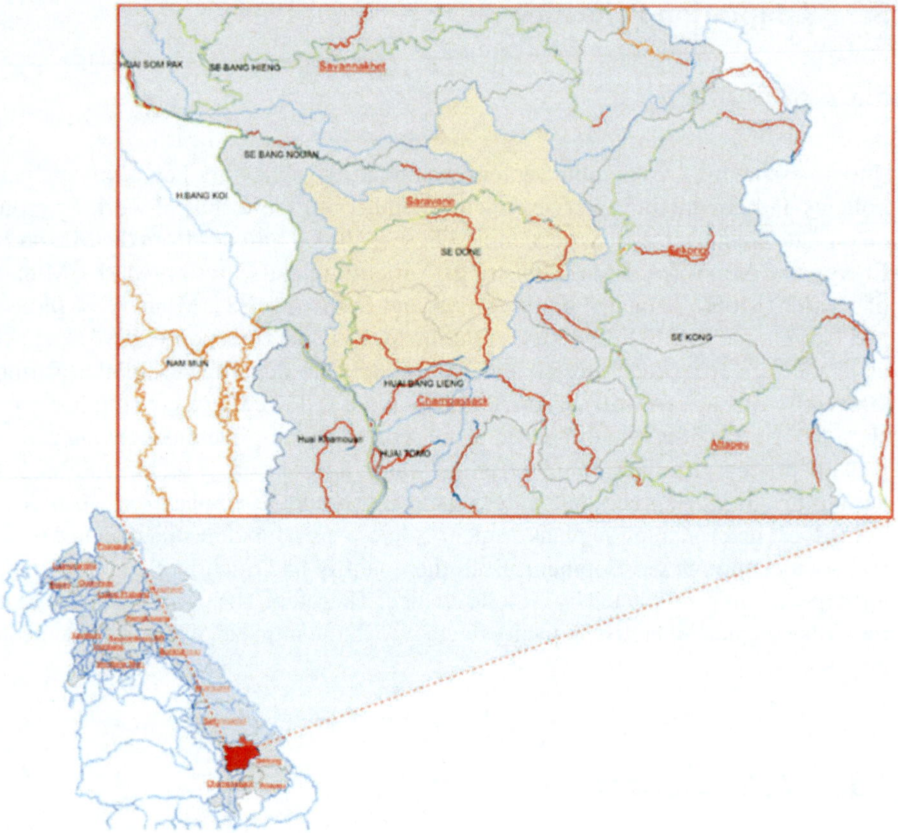

Fig. 5.9 Location of the proposed pilot in Lao PDR

would allow for advanced response to those impacts. Therefore, strengthening the water resource information system is critical.

The proposed pilot is expected to enhance the coping capacity of stakeholders to be better prepared to respond to floods. The provision of early warning information requires (i) modelling and vulnerability impact assessment for climate change in the river basin, (ii) development and implementation of early flood warning system, (iii) capacity building of technical staff, and (iv) dissemination of information via media, movies, posters, etc.

Agencies at the provincial level should take the lead in this pilot, with district governors and villagers playing a critical role, together with academics from the National University of Laos. The involvement of national agencies such as the Department of Meteorology and Hydrology, Department of Water Resources, Department of Housing and Urban Planning, Department of Irrigation, National Centre of Environmental Health and Water Supply, Department of Energy Policy and Planning, Department of Agriculture, Department of Energy Management, among others is vital. Technical support could also be sought from various regional institutions.

Co-financing could be pursued from donors such as the Asian Development Bank, World Bank, and Global Environment Facility, among others.

5.4 Concluding Remarks

The Climate Change Action Plan of Lao PDR (2013–2020), which draws on the experience of previous climate actions focuses on strengthening of institutional and human resource capacities, climate change adaptation, climate change mitigation as well as education and public awareness on climate change. The Action Plan is supplemented with the Guidelines on Ecosystem-Based Adaptation Practices to guide vulnerability assessment and selection of adaptation options at the provincial level in the country. The assessment of vulnerable sectors was based on reviewing key climate change policies as well as input from stakeholders such as government agencies, academia, and NGOs. The proposed local level pilot is the Se Done Basin in southern Lao PDR, to enhance the coping capacity of farmers and other stakeholders to be better prepared to respond to floods through the provision of early warning information.

Acknowledgements The author acknowledges the ASEAN-India Green Fund for supporting the Project on "Enhancing Local Level Climate Change Adaptation in Southeast Asia" (Phase 1). This chapter draws partly on findings of the Project, specifically the local level needs assessment and scoping exercise conducted in 2015.

References

ADPC (2010) Mainstreaming disaster risk reduction. Asian Disaster Preparedness Centre, Bangkok, Thailand

DDMCC (2013a) Climate change action plan of Lao PDR for 2013–2020. Ministry of natural resources and environment. Vientiane, Lao PDR

DDMCC (2013b) Technology needs assessment report "climate change adaptation". Ministry of natural resources and environment. Vientiane, Lao PDR

Fumagalli M (2020) Luang Prabang: Climate change and rapid development. Cities 97:102549. https://doi.org/10.1016/j.cities.2019.102549

IPCC (2007) Climate change 2007: synthesis report. An assessment of the intergovernmental panel on climate change. Cambridge University Press, United Kingdom

Kaisorn T (2015) DDMCC forum for implementation 5th year plan (2011–2015) and next 5th year plan (2016–2020) for natural resources and environment sector. Vientiane, Lao PDR

Lao PDR (2005) Forest strategy and action plan 2020. Ministry of agriculture and forestry. Vientiane, Lao PDR

Lao PDR (2011) The seventh five-year national socio-economic development plan (2011–2015). Ministry of planning and investment. Vientiane, Lao PDR

Lao PDR (2012) Lao census 2012, Lao Statistics Bureau. Ministry of planning and investment. Vientiane, Lao PDR

Lao PDR (2013) Second national communication on climate change (SNC). Ministry of natural resources and environment. Vientiane, Lao PDR

Lao PDR (2015) Natural resources and environment vision 2030. Ministry of natural resources and environment. Vientiane, Lao PDR

Lao Statistics Bureau (2012) Lao statistical yearbook-2012. Ministry of planning and investment. Vientiane, Lao PDR

MAF (2015) Forest cover assessment 2015. Department of forestry. Vientiane, Lao PDR

Ministry of Agriculture and Forestry (2005) Report on the assessment of forest cover and land use during 1982, 1992, 2002. Department of forestry. Vientiane, Lao PDR

Ministry of Energy and Mines (2008) Annual mine report 2008. Department of mines. Vientiane, Lao PDR

MRC (2010) Impact of climate change and development on Mekong flow regimes first assessment. MRC technical report 29. Vientiane, Lao PDR

Rizaldi B, Rakhman A, Faqih A (2013) Climate change vulnerability and adaptation needs in Lao PDR

Science, Technology and Environment Agency (STEA) (2000) First national communication on climate change. Vientiane, Lao PDR

Timmermann A, Oberhuber J, Bacher A, Esch M, Latif M, Roeckner E (2009) Increased El Nino frequency in a climate model forced by future greenhouse warming. Nature 398:694–697

World Bank (2010) Natural hazards, unnatural disasters the economic effective prevention. The world bank. Washington DC, USA

Chapter 6
Climate Change Adaptation in Malaysia

Joy Jacqueline Pereira and Mohd Khairul Zain

Abstract The National Policy on Climate Change promotes balanced adaptation and mitigation to achieve climate-resilient development. Notwithstanding, actions on climate change adaptation in Malaysia appear to lag behind climate change mitigation. The existing environmental management plans, national conservation strategies, disaster management plans as well as sector-based programs such as water, agriculture, forestry, transportation and fisheries, among others have elements of climate change adaptation. However, a specific national strategy for climate change adaptation is not yet been formulated nor is there a National Adaptation Plan developed to drive comprehensive and systematic action. On the research front, there is great uncertainty associated with the downscaling of regional climate data, and this calls for judicious use of climate projections for high investment decisions. Novel approaches are required for dealing with the anticipated impacts of climate change at the local level. A local level pilot is proposed in Kota Kinabalu, Sabah. The pilot is intended to address climate change risks associated with biomass and biofuel production, waste management, burden of diseases and coastal adaptation as well as institutional capacity, technology and resource challenges.

Keywords Climate change · Vulnerability · Hazard · Disaster risk · Malaysia

6.1 Introduction

Malaysia comprises two regions that are separated by the South China Sea: Peninsular Malaysia, and East Malaysia that is located on Borneo Island. The population of Malaysia was estimated to be about 32.4 million in 2018 with an annual growth rate of 1.1% (DOSM, 2018). Peninsular Malaysia is dominated by a central mountainous spine extending from the north to the south, known as the Titiwangsa Range, about 480 km in length and 900–2,100 m above sea level. The topography of East Malaysia is generally mountainous in the northwest, where the highest peak in Southeast Asia, Mount Kinabalu at 4,095 m above sea level, is located.

J. J. Pereira (✉) · M. K. Zain
Universiti Kebangsaan Malaysia, Bangi, Malaysia
e-mail: joy@ukm.edu.my

J. J. Pereira et al. (eds.), *Climate Change Adaptation in Southeast Asia*,
Disaster Risk Reduction, https://doi.org/10.1007/978-981-16-6088-7_6

Malaysia experiences relatively uniform diurnal variations of temperatures throughout the year where the daily mean temperature ranges between 26 °C and 28 °C (MESTECC 2018). The wind over the country is generally light and variable, with some uniform periodic changes in flow patterns. Northeasterly winds persist during the northeast monsoon from November to March, and the southwesterly winds prevail during the southwest monsoon from May to September. These monsoons are separated by two shorter inter-monsoon periods. The weather along coastlines is affected by convective rain and rainfall distribution is greatly influenced by topography and monsoon winds. Malaysia receives abundant rainfall with an average annual rainfall ranging from about 2,000–4,000 mm (MESTECC 2018). During the northeast monsoon, the east coast of Peninsular Malaysia and parts of East Malaysia experience heavy rainfall lasting from 1 to 3 days, which, sometimes, results in severe floods. The country generally receives about 6 hours of direct sunlight every day.

Climate projections for the country have provided an insight into future conditions with respect to temperature, rainfall, stream flow, evapotranspiration and soil moisture over time slices of up to 2050 (MESTECC 2018). However, there is great uncertainty associated with the downscaling of regional climate data using a limited number of general circulation model (GCM), which is further compounded by knowledge gaps due to poor observational records and understanding of regional climate phenomena (Tangang et al. 2012; Cruz et al. 2017; Galavi et al. 2019; Kong et al. 2019; Tan et al. 2019). A more coordinated approach is required to enhance research for generating more accurate projections that will be of use to policy- and decision-makers in the country. Given the current shortcoming, current climate projections for the country have to be applied judiciously, particularly in relation to high investment decisions. In addition, other approaches have to be developed for identifying potential climate change impacts at the local level so that appropriate adaptation measures can be identified.

6.2 National Perspective

6.2.1 National Arrangements and Institutions

Malaysia is taking various measures to address the challenges of climate change. The mainstreaming of climate change is reflected in the objectives and strategies articulated in national development policies. The current development narrative for Malaysia is embodied in the Shared Prosperity Vision with the aim of providing a decent standard of living for all Malaysians by 2030 (GoM 2019). Vision 2030 has three objectives that are driven by 15 guiding principles, and 8 enablers covering 7 strategic thrusts, and 15 key economic growth activities. Climate change comes under the Strategic Thrust on Social Capital. Among the key economic growth activities linked to climate change are renewable energy, green economy, logistics, transportation and sustainable mobility. A total of 10 economic indicators and 8 social indicators have been identified to measure progress; one of these is the environmental and climate change index. The Shared Prosperity Vision 2030 builds on previous policies with climate change elements in the country.

The National Policy on Climate Change (2009) promotes balanced adaptation and mitigation to achieve climate-resilient development. This is supported by the The National Policy on Climate Change (2009) that focuses to enhance usage and diffusion of green technology to help in the reduction of carbon emission in the country. Malaysia has made great strides in climate change mitigation with a national aspiration to reduce GHG emission intensity of Gross Domestic Product voluntarily by up to 40% of 2005 levels by 2020. The Eleventh Malaysia Plan, covering the period from 2016 to 2020, has articulated the need for reduced reliance on petroleum products (MESTECC 2018). This is a serious commitment given the high dependency of the country on natural gas, petroleum products and coal. Efforts are being made to expand the renewable energy market, strengthen energy efficiency and explore new sources of energy such as wind and geothermal energy as part of the renewable energy mix (MESTECC 2018).

Climate change adaptation is also embedded in specific sectors through existing environmental management plans, national conservation strategies, disaster management plans as well as programs within other sectors such as water, agriculture, forestry, transportation and fisheries, among others. (Pereira et al. 2012; Zen et al. 2016). Examples include the National Water Resources Policy of 2012 and the National Agro-food Policy adopted in 2011, with the aim of ensuring adequate food supply and safety. The Green Technology Master Plan (2017–2030) facilitates the mainstreaming of green technology in development planning with a focus on six key sectors: energy, transportation, manufacturing, building, waste and water. Initiatives in the policy platform of the country have provided an impetus to mobilize the climate change agenda in the science platform. Notwithstanding this, more needs to be done to enhance adaptation measures in the country, both in the policy and science domains.

Malaysia is committed to addressing the existential threat of climate change at the global level. The country is a signatory to the United Nations Framework Convention on Climate Change (UNFCCC) with the Ministry of Environment and Water (KASA) currently serving as the National Focal Point. Malaysia is also a signatory to the Paris Agreement. The National Steering Committee on Climate Change was established in January 1995 after the country ratified the UNFCCC in 1994. Through KASA, the country also plays an active role in the Intergovernmental Panel on Climate Change (IPCC), with support from the Malaysian Meteorological Department and other partners. Malaysia submitted the Initial National Communication (INC) to the UNFCCC in 2000, the Second National Communication (NC2) in 2011 and the Third National Communication (NC3) in 2018 (MOSTE 2000; MESTECC 2018). The process of preparing the National Communications has served as a means of bringing together policy-makers and researchers from research institutes and universities, to view climate change in its broad multidisciplinary context.

6.2.2 Vulnerable Sectors and Regions

Disaster Risk Reduction: Climate change and disaster risk reduction are generally implemented in different policy domains involving different custodians. The National Disaster Management Agency (NADMA) Malaysia is the country's National Focal Point to the United Nations Office on Disaster Risk Reduction (UNDRR) and reports progress on managing disasters to the Sendai Framework for Disaster Risk Reduction via the Sendai Monitor. The Ministry of Environment and Water (KASA) is the National Focal Point to the UNFCCC and IPCC. In Malaysia, there has been a conscious effort to reconcile these components through the National Policy on Climate Change (Pereira et al. 2012). This is critical to promote climate change adaptation and disaster risk reduction as a priority at national, state and local authority levels with a strong institutional basis for implementation. Adaptation measures should be based on the provision of appropriate information regarding vulnerable and exposed areas so that it will also serve to strengthen spatial planning at the local level . Transdisciplinary research should be encouraged to identify the potential impacts and cascading effects associated with both fast- and slow-onset hazards (Pereira et al. 2019a; Yahaya et al. 2016). Adaptive capacity should be improved at the local level through multi-stakeholder partnerships, community-based actions and engagements, policy-relevant knowledge generation as well as education and awareness to build a culture of safety and enhance resilience. Ongoing programs led by operational agencies in the country such as the Meteorology Department of Malaysia, Minerals and Geoscience Department of Malaysia, and Drainage and Irrigation Department Malaysia among others, have the potential to strengthen risk management associated with climate-related hazards and emerging events associated with climate change. With the involvement of city managers and local level stakeholders, these initiatives could strengthen the adaptive capacity in cities, towns and rural settlements, particularly to enhance land-use planning and decision-making for major structural investment.

Water: Effective water resources management is a prerequisite for a successful Economic Transformation Plan in the country as water is essential for human settlements and its need cuts across all industrial sectors including agriculture (Zakaria et al. 2011). Water resources are highly sensitive to climate change and poor management could result in floods, droughts and water pollution. There is a wide range of adaptation measures that have been applied for the water sector but more needs to be done at the local level. National policies were found to facilitate the adaptive capacity of critical water institutions but this is not supported by State and basin level plans, which have to be reviewed for this purpose (ASM 2018). The costs of adaptation can be lowered by strengthening the adaptive capacity of water institutions in the country. This measure can be considered a "no-regret" option. Stakeholders from the water sector are now focused on identifying goals and aspirations for the water sector to develop immediate, medium- and long-term programs to transform the water sector. The transformation plan involves Integrated Water Resources

Management (IWRM), which will be implemented through strategic interventions to empower people, strengthen governance and enhance finance, information and infrastructure (ASM 2016). There is great potential to promote nation-wide adaptation to climate change in the country by enhancing the implementation of IWRM at the river basin scale.

Agriculture: Important crops in Malaysia include rice as well as commodities such as oil palm, rubber and cocoa. Climate change is expected to bring with it warmer temperatures and drier or wetter conditions where the actual rise will depend on the degree of emission reductions achieved in the next few decades (IPCC 2018). Climate change impacts on the agriculture sector in the country have to be studied in the context of comprehensive strategies for climate change adaptation and mitigation (MESTECC 2018). These include, among others, revitalization of the sector as a priority to ensure food security and increased self-reliance without compromising the environment; improved water efficiency; enhanced community-based activities and increased substantial investments in agriculture to promote greater engagement and productivity (Devendra and Leng 2011). With respect to mitigation, the sector could benefit by exploring the potential to enhance carbon sequestration and reduce emissions and its synergies with climate change adaptation.

Health: Continuous monitoring and prevention are conducted under the public health system to address diseases influenced by climates such as dengue, malaria, food and water-borne diseases as well as chronic non-communicable diseases sensitive to air pollution and heat. Climate change is expected to bring new challenges to public health service. It is estimated that about 7–8% of the public healthcare facilities in the country are exposed to floods, and the number affected is likely to increase in the 2030–2050 period (MESTECC 2018). Some of the facilities are located near coastal areas and their exposure to sea-level rise has to be investigated. With respect to dengue, more effective tools are required for better adaptation response to the changing climate (Abd Samad et al. 2012). This requires improving methods for mosquito control, increasing knowledge of dengue transmission, enhancing understanding of the influence of climatic factors and linkages to environmental, immunity and socio-behavioral factors. Malaria cases are restricted to rural areas making monitoring and control measures a logistic challenge. Research is required to increase the understanding of several aspects of the health sector. These include linkages between air pollution and the impact of climatic factors on morbidity and mortality, the impact of rising temperatures on temperature-sensitive illnesses and the health impact of weather extremes. The capacity for quantitative and qualitative vulnerability assessment for the health sector at the local level should be strengthened. As health impacts of climate change are also influenced by other non-health factors such as socioeconomic status, population growth and migration, land use and landscape changes, environmental factors and ecosystem services, local level assessments will be more effective if they are spatially contextualized.

Forestry: The forestry sector focuses on measures that promote sustainable forest management (Wan Mohd 2011). Research is required to maintain this practice in view of emerging challenges. Forest management practices have to meet the traditional consignment for wood in addition to the demand for non-wood products, protection of ecosystem services, provision of social and recreational needs as well as management of environment and biodiversity issues. Climate change is now an emerging challenge where strategic actions involving multiple stakeholders are required to promote emission reduction. Issues related to Reduced Emissions from Deforestation and Forest Degradation Plus (REDD+), which extends to conservation, enhancement of forest carbon stock and sustainable management of forest or sustainable forestry practices, require further attention.

Ecosystems and Biodiversity: Malaysia has the necessary framework, laws and regulations that have addressed conservation requirements. These can be readily employed to address climate change concerns related to ecosystems and biodiversity. The country has about 56% of its land under permanent forest and 78% under tree canopy (MESTECC 2018). Ecosystem services and biodiversity are affected by climate change but strong and resilient ecosystems are critical for mitigation. A well-functioning ecosystem can be a very cost-effective response to climate change and constitutes a good investment. Rainforest ecosystems serve as resources for the timber industry and host diverse genetic resources that represent an irreplaceable information database. For example, over 15,000 species of flowering plants and 185,000 of animal species, which account for about 9 and 16% of the world's total are found in Malaysia (Wan Mohd 2011). Research is required to identify the most vulnerable ecosystems in the country. The capacity of these ecosystems to adapt to climate change, serve as carbon sequesters and influence local climate, has to be investigated in conjunction with mitigation strategies that reduce emissions. Linkages to other forms of stressors as well as cross-sector relationships at the local level should also be assessed. These findings should serve as the basis for formulating strategic plans based on the ecosystems approach, appropriate for both local and regional levels, to meet the challenges of a changing climate.

Energy and Industry: The energy outlook for the country requires careful and strategic policy intervention and technological innovation. Many initiatives have commenced on energy efficiency and renewable energy. Further investments in research and development, and technology diffusion have the capacity to reduce imminent mitigation costs and contribute to an array of appropriate technology-based interventions to meet the policy goal that has been set for energy-related sectors in the country (MESTECC 2018). Climate change mitigation responses under investigation include carbon capture and storage, and green technologies as well as a comprehensive ecological industrial policy including capacity building, technology and innovation, to transform industry and society toward sustainable production and development.

Capacity Building and Awareness: The Malaysian Government has identified several areas that require capacity building to enable the country to advance in dealing with the impacts of climate change (Table 6.1). These include the capacity to generate appropriate information on climate drivers such as temperature, precipitation and sea-level rise as well as hazards with the potential to impact a variety of sectors. In addition, there is a need to increase the awareness of stakeholders from multiple

Table 6.1 Key gaps and proposed improvements identified by the Malaysian Government to advance climate change adaptation in the country. *Source* MESTECC (2018)

Area	Gaps	Improvement plan
Flood	Lack of projected flood maps covering all the flood-prone basins	Establish comprehensive flood maps covering all flood-prone basins in Malaysia
Dry spell	Analysis of projected dry spells for vulnerability assessments need to be based on seasonal rather than annual scale, and with high spatial resolution	Develop seasonal projected dry spell maps that are suitable for vulnerability assessment at the sub-basin and local scales
	Insufficient priority and detailed analysis of impact of dry spell on dams and reservoirs	Conduct studies on the impact of climate change on dams for electricity generation and water supply
Sea level rise	Insufficient coastal hydrodynamic simulation that incorporates sea level rise impacts for vulnerable coastal regions in Malaysia	Develop high-resolution coastal inundation maps based on coastal hydrodynamic simulation that incorporates sea level rise for 20-year time intervals up to 2100, and carry out detailed coastal erosion and sedimentation studies to plan comprehensive adaptation measures
Integrated hazards	The combined impacts of sea level rise, storm surges, abnormally high tides and high rainfall which could lead to severe flooding are not well understood	Develop models that integrate sea-level rise, storm surges, abnormally high tides and high rainfall for early warning and assessment of future scenarios
	Inadequate information of assets in vulnerable areas for each sector	Develop a comprehensive database for each sector
Capacity building and awareness	Insufficient capacity to conduct impact assessment in all key sectors	Develop the required capacity in each key sector for V&A analysis
	Lack of awareness and understanding of the full chain of implications of climate change impacts by key stakeholders in all relevant sectors	i. Conduct detailed studies on the full chain impacts of climate change for all sectors
		ii. Conduct awareness and capacity building programs to targeted stakeholders

sectors at federal, state and local levels regarding the potential impacts on climate change that may affect their health, well-being, livelihood and investments. These gaps could be addressed effectively through collaborations involving government, academia, civil society, non-government organizations and the private sector.

6.2.3 Ongoing Initiative: Disaster Resilient Kuala Lumpur

In the advent of climate change, resilience has become an important goal for cities. In this context, resilience refers to the ability of a city to avoid or recover from an adverse event. Kuala Lumpur is the primary business hub of Malaysia, and it is important that the city has the capacity to adapt and expand despite the stresses and shocks that may be experienced as the climate changes. An incredible level of coordination across multiple levels of stakeholders and sectors is required to make Kuala Lumpur resilient. The City Hall of Kuala Lumpur and related stakeholders must be supported with adequate scientific input to make informed decisions. Such information also includes aspects of risk, vulnerability and exposure associated with climate change. Engineering solutions, as well as community actions, have been taken to cope with frequent flash floods and landslides that occur in the city (Bhuiyan et al. 2018; Rustam and Raja 2013). While not serious enough to be considered a disaster of national scale, these events are disruptive and pose an inconvenience to the community and commercial activities in the city. The availability of well-organized information is required to support decision-makers in the City Hall of Kuala Lumpur to better manage the risk of floods, landslides and other events such as sinkholes, strong winds, air pollution and urban heat (Pereira et al. 2019b).

This was addressed via the pilot project entitled "Disaster Resilience Cities: Forecasting Local Level Climate Extremes and Physical Hazards for Kuala Lumpur", also referred to as Disaster Resilient Kuala Lumpur, supported by the Newton-Ungku Omar Fund, which is a bilateral program with matching funds from the Governments of the United Kingdom and Malaysia. The pilot, jointly led by the University of Cambridge and Universiti Kebangsaan Malaysia's Southeast Asia Disaster Prevention Research Initiative (SEADPRI-UKM), involved 16 organizations from academia, government agencies, non-government organizations and the private sector from both countries. Products from the pilot include the first-ever 3-day street-level forecast of rainfall, wind speed and temperature for the city, to strengthen early warning for flash floods and landslides, as well as a system for atmospheric hazard forecasting, which are displayed in the Kuala Lumpur Multi-Hazard Platform (Pereira et al. 2019b). In addition, areas susceptible to flash floods and landslides have also been identified to facilitate targeted interventions to build community resilience and reduce increasing inequalities due to disaster events. The enhanced local level forecasting capability of the multi-hazard platform will support the City Hall of Kuala Lumpur in developing strategies that strengthen the institutional capacity to reduce the number of affected people, foster disaster resilience and enhance the ability to adapt to climate change.

Fig. 6.1 The Kuala Lumpur Multi-Hazard Platform developed by the University of Cambridge, Southeast Asia Disaster Prevention Research Institute (SEADPRI-UKM) and partners with support from the Newton Ungku Omar Fund, provides local level forecasts of climate extremes and physical hazards

The Kuala Lumpur Multi-Hazard Platform could be replicated in other cities in the region to support city managers with adequate scientific inputs to make informed decisions and build community resilience to disaster risks and climate change (Fig. 6.1). Severe and extreme weather events associated with climate change are projected to increase losses, challenging Governments and insurance systems worldwide. Communication, transfer and development of timely climate-related risk information to city managers are critical so that they are forewarned and better equipped to cope with the threats ahead.

6.3 Adaptation Priorities

6.3.1 Approach

The capacity for adaptation and mitigation is a challenge for Malaysia with respect to the availability of information, finance and societal values (MESTECC 2018). The lack of capacity is a major challenge to transform the country into a low-carbon society and undermines the effort for sustainable development. Progress on climate change adaptation appears to be slow compared with climate change mitigation. The country has not framed a national strategy for climate change adaptation or developed a National Adaptation Plan as advocated by the UNFCCC. This has hindered efforts to systematically enhance adaptive capacity, build resilience and integrate adaptation into policies, plans and strategies in a coherent manner into various sectors and at the relevant scale. A key shortcoming in Malaysia is the lack of an overall framework, developed in conjunction with local stakeholder input and involvement of scientists from multiple disciplines, as well as non-government organizations and the private sector.

There is a high level of uncertainty related to climate change models in the country (Tangang et al. 2012; Galavi et al. 2019). The conventional end-point approach to adaptation is not suitable for the country, especially at the river basin scale as uncertainties associated with climate models are far greater than those related to hydrological models. Practitioners at the global level have already noted that such an end-point approach has limited success in facilitating pragmatic impact assessment and adaptation planning (Alves et al. 2020; Toimil et al. 2020). Given that there is still a lot of uncertainty related to climate change models, a process-oriented starting point adaptation approach is more suitable for the country. The starting point approach draws on the best available science to identify low-regret adaptation actions, with periodic review of the situation as more scientific information becomes available. A strategic entry point for starting point adaptation in the country is via the water sector transformation plan, which is expected to commence in the 12th Malaysia Plan for the period 2021–2025. The IPCC risk framing approach facilitates the integration of both fast- and slow-onset climatic hazards (IPCC 2014; Pereira et al. 2019a). In conjunction with geographic context, best available science and stakeholder consultation, disaster resilience plans that integrate short-term risks due to climate variability and long-term risks associated with climate change could be developed (Fig. 6.2). The disaster resilience plans could serve as a basis for starting point adaptation at the local level.

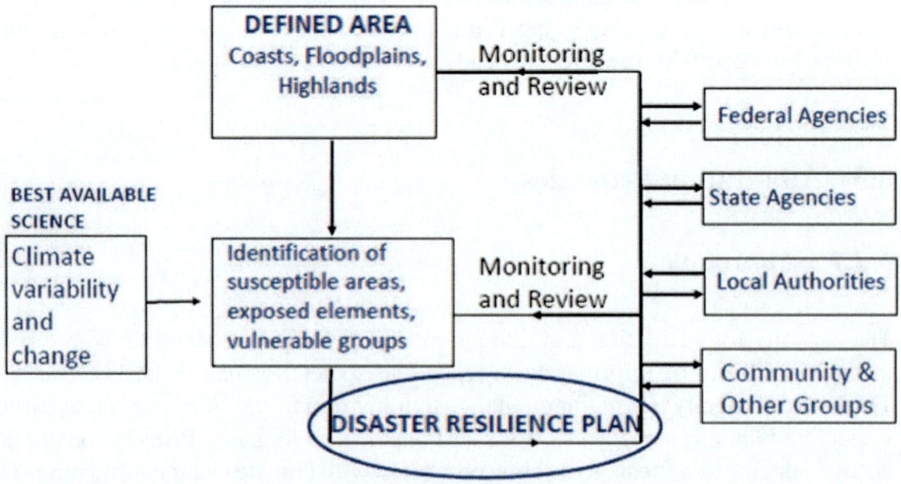

Fig. 6.2 Local level pilots that are contextualized to defined geographic areas that draw on the best available science with input from multiple stakeholders are useful to develop disaster resilience plans that integrate fast- and slow-onset climatic hazards

A series of consultative engagement with multiple stakeholders was held in 2015 to identify needs and priorities for local level climate change adaptation within the ASEAN-India Green Fund Project. The initiative culminated in a workshop involving 30 participants, primarily from government agencies, to discuss climate change adaptation challenges and identify several local level pilot initiatives. Four themes were prioritized during the workshop; these were (i) Climate Change Impacts on Biomass and Biofuel Production in Sabah; (ii) Risk Assessment of the Burden of Disease from Climate Change in Kuala Lumpur; (iii) Coastal Adaptation Strategies for Matang Mangrove in Perak and (iv) Adaptation Framework for Cities, Focusing on Waste Management in Kota Kinabalu, Sabah. Major challenges highlighted encompassed local level institutional capacity, technology and resources.

6.3.2 *Potential Local Level Pilot*

The City of Kota Kinabalu, capital of the State of Sabah was proposed as a local level pilot. All the four themes that were previously identified could be addressed in this locality, within the adaptation framework for cities. Aspects to be addressed include the development of adaptation strategies, including risk assessment of the burden of disease from climate change in the coastal capital city. Constraints to be tackled for biomass and biofuel production include inefficiency in collection and delivery of waste, absence of household waste separation as well as poor grid connectivity of supply. Recent investigation has revealed that waste disposal sites in Peninsular Malaysia could be affected by the sea level rise and exposed to other climate-related geohazards such as landslide and other cascading effects (Yahaya et.al. 2016). Such an investigation is also critical to provide baseline information to improve waste management in the City of Kota Kinabalu. Other needs include a feasibility study, transfer of technology, training for local people, a facilitative policy framework as well as a sustainable business plan. The State Secretary of Sabah and the Mayor of Kota Kinabalu could provide the institutional leadership for the pilot with technical support from Universiti Malaysia Sabah (UMS). The potential for co-financing could be explored from the Federal and State Governments, the private sector and other international sources.

6.4 Concluding Remarks

Climate change is addressed in several national development plans and policies in Malaysia. The National Policy on Climate Change promotes balanced adaptation and mitigation to achieve climate-resilient development. Climate change adaptation is incorporated in specific sectors including environmental management, conservation, disaster management as well as water, agriculture, forestry, transportation and fisheries, among others. However, the country does not have a national strategy for climate change adaptation or a National Adaptation Plan to systematically enhance adaptive capacity, build resilience and coherently integrate adaptation into various sectors and at the relevant scale, in conjunction with stakeholder engagement. Actions on climate change adaptation appear to lag behind climate change mitigation.

On the research front, there is great uncertainty associated with the downscaling of regional climate data, and this calls for judicious use of climate projections for high investment decisions. Novel approaches including capacity building and awareness raising are required for addressing future impacts at the local level so that appropriate adaptation measures can be identified. An example is the Kuala Lumpur Multi-Hazard Platform developed for the City Hall of Kuala Lumpur, to facilitate the development of strategies that strengthen institutional capacity, foster disaster resilience and enhance the ability to adapt to climate change. Stakeholder engagement conducted in Malaysia resulted in the proposal of a local level pilot in the City of Kota Kinabalu, Sabah. The pilot is intended to address climate change risks associated with biomass and biofuel production, waste management, the burden of diseases and coastal adaptation as well as institutional capacity, technology and resource challenges. With institutional leadership from the State of Sabah and City of Kota Kinabalu and technical support from Universiti Malaysia Sabah (UMS), the potential for co-financing for the pilot from multiple sources could be explored.

Acknowledgements The authors acknowledge the ASEAN-India Green Fund for supporting the Project on "Enhancing Local Level Climate Change Adaptation in Southeast Asia" (Phase 1). This chapter draws partly on findings of the Project, specifically the assessment and scoping conducted in 2015 to identify the local level pilot. Mr. Jaya Singam Rajoo, Mr. Muhamad Ridzwan Bin Ali, Mr. Yusmazy Bin Md Yusup and Dr. Gary Thesiera from the former Ministry of Energy, Science, Technology, Environment and Climate Change (MESTECC) are thanked for their support while Dr. Lian Kok Fei is acknowledged for his role in initiating the ASEAN-India Green Fund Project at the ASEAN Working Group on Climate Change (AWGCC).

References

Abd Samad BH, Choy EA, Othman H, Sulaiman LH, Sahani M, Zailani MH, Nik Hassan NMN, Zainon NA, Ramachandran CP, Hod R, Zainuddin MA (2012) Health impacts of climate change. ASM series on climate change. Academy of Sciences Malaysia ISBN 978–983–9445–79–4

Alves F, Filho WL, Casaleiro P, Nagy GJ, Diaz H, Al-Amin AQ, Guerra JBSOA, Hurlbert M, Farooq H, Klavins M, Saroar M, Lorencova EK, Suresh J, Soares A, Morgado F, O'Hare P, Wolf F, Azeiteiro UM (2020) Climate change policies and agendas: Facing implementation challenges and guiding responses. Environ Sci Policy 104:190–198. https://doi.org/10.1016/j.envsci.2019.12.001

ASM (2016) National integrated water resources management plan: strategies and road map, vol 1. Academy of Sciences Malaysia Advisory Report, Kuala Lumpur, p 135

ASM (2018) Climate change: adaptive capacity attributes of selected water-related policies in Malaysia. Academy of Sciences Malaysia Advisory Report, Kuala Lumpur, p 47

Bhuiyan TR, Reza MIH, Choy EA, Pereira JJ (2018) Facts and trends of urban exposure to flash flood: a case of Kuala Lumpur city. Improving flood management, prediction and monitoring, p 79–90. https://doi.org/10.1108/S2040-726220180000020016

Cruz FT, Narisma GT, Dado JB, Singhruck P, Tangang F, Linarka UA, Wati T, Juneng L, Phan-Van T, Ngo-Duc T (2017) Sensitivity of temperature to physical parameterization schemes of RegCM4 over the CORDEX-Southeast Asia region. Int J Climatol 37:5139–5153. https://doi.org/10.1002/joc.5151

DOSM (Department of Statistic Malaysia) (2018) Current population estimates for Malaysia 2017–2018. https://www.dosm.gov.my/v1/index.php?r=column/pdfPrev&id=T0lmSHlwOXJ1aWlzcjBlam9ObUZpdz09. Accessed 5 August 2018

Devendra C, Leng RA (2011) Feed resources for animals in Asia: issues, strategies for use, intensification and integration for increased productivity. Asian- Aust. J. Anim. Sci. 24(3):303–321. https://doi.org/10.5713/ajas.2011.r.05

Galavi H, Kamal MR, Mirzaei M, Ebrahimian M (2019) Assessing the contribution of different uncertainty sources in streamflow projections. Theoret Appl Climatol 137(1–2):1289–1303. https://doi.org/10.1007/s00704-018-2669-0

GoM (Government of Malaysia) (2019) Shared prosperity vision 2030: restructuring the priorities of Malaysia's development-ministry of economic affairs Malaysia. Percetakan Nasional Malaysia Berhad, Kuala Lumpur

IPCC (2014). Climate change 2014: synthesis report. contribution of working groups I, II and III to the fifth assessment report of the intergovernmental panel on climate change. IPCC, Geneva, p 151

IPCC (2018) Summary for policymakers. In: Masson-Delmotte V, Zhai P, Pörtner H-O, Roberts D, Skea J, PR Shukla A, Pirani W, Moufouma-Okia, Péan C, Pidcock R, Connors S, Matthews JBR, Chen Y, Zhou X, Gomis MI, Lonnoy E, Maycock T, Tignor M, Waterfield T (eds.), Global warming of 1.5°C. An IPCC special report on the impacts of global warming of 1.5°C above pre-industrial levels and related global greenhouse gas emission pathways, in the context of strengthening the global response to the threat of climate change, sustainable development, and efforts to eradicate poverty. World Meteorological Organization, Geneva, Switzerland, p 32

Kong SSK, Sentian J, Chuang MT, Ooi MCG, Chee FP, Chang JHW (2019) Simulation analysis of the future surface ozone in Malaysian region under representative concentration pathways (RCPs) emission scenarios. Int J Environ Sci Technol 16(11):7357–7374. https://doi.org/10.1007/s13762-019-02322-6

MESTECC (2018) Malaysia third national communication (NC3) and second biennial update report to the UNFCCC. Ministry of energy, science, technology, environment and climate change. Putrajaya, ISBN: 978–967–13297–4–0

MOSTE (2000) Malaysia initial national communication. Ministry of science, technology and environment. Kuala Lumpur

Pereira JJ, Komoo I, Tan CT, Che Moin U, Lian KF (2012) Climate change and disaster risk reduction. Academy of sciences Malaysia series on climate change. Kuala Lumpur, p 16

Pereira JJ, Pulhin JM, Nyda C, Tran DT, Khadijah S (2019a) Appraising slow onset hazards for loss and damage: case studies in Southeast Asia. APN Scie Bullet 9(1):45–51. https://doi.org/10.30852/sb.2019.720

Pereira JJ, Nurfashareena M, Sian LC, Aziz S, Hunt JCR (2019b) Making cities disaster resilient in a changing climate: the case of Kuala Lumpur. IRDR working paper series, Malaysia, p 1–17. https://doi.org/10.24948/2019.05

Rustam KZ, Raja NRA (2013) Risk communications: flood-prone communities of Kuala Lumpur. Procedia Environ Sci 17:880–888. https://doi.org/10.1016/j.proenv.2013.02.106

Tan ML, Juneng L, Tangang FT, Chan NW, Ngai ST (2019) Future hydro-meteorological drought of the Johor River Basin, Malaysia, based on CORDEX-SEA projections. Hydrol Sci J 64(8):921–933. https://doi.org/10.1080/02626667.2019.1612901

Tangang FT, Juneng L, Salimun E, Sei KM, Le LJ, Muhamad H (2012) Climate change and variability over Malaysia: gaps in science and research information. Sains Malaysiana 41(11):1355–1366

Toimil A, Losad IJ, Nicholls RJ, Dalrymple RA, Stive MJF (2020) Addressing the challenges of climate change risks and adaptation in coastal areas: a review. Coast Eng 156:103611. https://doi.org/10.1016/j.coastaleng.2019.103611

Wan Mohd WR (2011) Climate change and biodiversity in Malaysia: research, development and policy issues in sustainable forestry: Workshop on climate change & biodiversity-mobilizing the research agenda. Universiti Kebangsaan Malaysia (UKM), Bangi, Selangor, Malaysia, 13–14 December

Yahaya NS, Lim CS, Taha MR, Pereira JJ (2016) Exposure of municipal solid waste disposal sites to climate related geohazards: case study of Selangor. Bull Geol Soc Malays 62:57–63. https://doi.org/10.7186/bgsm62201608

Zakaria S, Jin L, Iyer L (2011) Water resources and climate change. In: Proceedings of the national conference on the impact of climate change on water resources and their consequences to major economic sectors, 25 July 2011, Ministry of natural resources and environment (NRE). Putrajaya, Malaysia, p 15–39

Zen IS, Abul Quasem AA, Zeeda Fatimah M, Mohd Hamdan A, Syed Ahmad Iskandar SA, Fera Fizani AF (2016) Mainstreaming climate adaptation and mitigation policy in Malaysia: case study of Melaka City. International symposium on sustainable development management (October)

Chapter 7
Climate Change Adaptation in Myanmar

Mohd Khairul Zain, Mitsuko Shikada Otsuyama, and Rajib Shaw

Abstract Myanmar is exposed to multiple natural hazards including cyclones, earthquakes, floods, droughts, landslides, tsunamis, and fires. The western coast of Myanmar is subject to frequent tropical storms and cyclones during monsoon seasons. Socioeconomic sectors related to adaptation in Myanmar include agriculture, forest, biodiversity, coastal zones, public health, water resources, and education. The agriculture sector in Myanmar, particularly agricultural production and food security, is very vulnerable to the negative impacts of climate change. Potential local level pilots have been identified in the dry zone and coastal region.

Keywords Agriculture · Climate change · Vulnerability · Disaster risks · Myanmar

7.1 Introduction

The Republic of the Union of Myanmar is located in Southeast Asia with a total land area of about 676,578 km^2 and it borders India and Bangladesh in the west, China in the North, and Lao PDR and Thailand in the east; it faces the Bay of Bengal and the Andaman Sea in the south and the southwest. According to the World Bank country data, the population of Myanmar is estimated at 54.6 million. About 66% of the population is rural-based, while 33% live in urban areas (Department of Meteorology and Hydrology 2012). The population density is highest along the Ayeyarwady Basin. The basin is located between the Rakhine Yoma and the Shan Plateau, encompassing Sagaing, Magway and Mandalay Regions in the Central Belt. The humanitarian and political crisis in Myanmar has become more intense since 2016 (South 2020). This is overshadowing, to some extent, the climate change issues the country faces.

M. K. Zain
Universiti Kebangsaan Malaysia, Bangi, Malaysia

M. S. Otsuyama (✉)
SEEDS Asia, Kobe, Japan
e-mail: mitsuko.otsuyama@seedsasia.org

R. Shaw
Keio University, Fujisawa, Japan

J. J. Pereira et al. (eds.), *Climate Change Adaptation in Southeast Asia*,
Disaster Risk Reduction, https://doi.org/10.1007/978-981-16-6088-7_7

The climate in Myanmar ranges from tropical to subtropical with three seasons. These are the mainly hot and dry inter-monsoonal season from mid-February to mid-May; followed by the rainy southwest monsoon from mid-May to late October; and finally the cool, relatively dry northeast monsoon from late October to mid-February. The mean annual rainfall is the lowest in the Central Dry Zone (500–1000 mm per year). The rainfall increases in the eastern and northern hilly regions and is the highest in the Southern and Rakhine Coastal regions (2500–5500 mm). The mean temperature ranges from 32 °C in the coastal and delta areas to 21 °C in the northern lowlands.

Myanmar is exposed to multiple natural hazards including cyclones, earthquakes, floods, droughts, landslides, tsunamis, and fires (Union of Myanmar 2009). Fire is the most frequent disaster in Myanmar and accounts for 71% of the disasters within the country, followed by storms (11%) and floods (10%), with the remaining 8% of disasters which include earthquakes, landslides, etc. The western coast of Myanmar always experiences tropical storms and cyclones, particularly during the monsoon season. Myanmar has been periodically hit by natural disasters, notably Cyclone Nargis in 2008 which caused about 138,373 deaths and had many people missing, with damage to property at approximately USD 4.1 billion (Union of Myanmar 2009).

7.2 National Perspective

7.2.1 National Arrangements and Institutions

The Government of Myanmar is taking action to tackle environmental stress on socioeconomic sectors. Ministries have laid out policies, objectives, and strategies that are indirectly related to climate change adaptation. In 1990, the National Commission for Environmental Affairs (NCEA) was established with the aim of managing and coordinating environmental matters in the country, and acting as the national focal point for environmental relations with other countries and international organizations. The NCEA was reformed in 2011 as a focal organization for the environment sector, and finally, it was renamed the Ministry of Environmental Conservation and Forestry (MOECAF) in 2011 (from Ministry of Forestry) to ensure that environmental affairs are considered in development projects. There are several treaties/agreements addressing climate change.

The United Nations Framework Convention on Climate Change (UNFCCC) is an international environmental treaty established at the Earth Summit in Rio de Janeiro. Myanmar ratified UNFCCC on 25 November 1994 as a non-Annex I Party. Article 12.5 of the UNFCCC requires non-Annex I Parties to make their initial national communications (INC). Therefore, the INC project was implemented in 2008; it was financially supported by the Global Environment Facility (GEF)/United Nations Environment Programme (UNEP), and it presents the findings of the thematic groups, and suggestions and recommendations for the Myanmar Government to undertake designing measures addressing the climate change issues. Myanmar ratified the Kyoto Protocol in 2003. More recently the country has submitted its Nationally Determined Contribution after ratifying the Paris Agreement. Myanmar's National Adaptation Program of Action (NAPA) (2012) identifies and communicates priority activities to address Myanmar's immediate and urgent adaptation needs for climate change. Thirty-two priority adaptation projects were identified for the following sectors: agriculture, early warning system, forest, public health, water resources, coastal zones, energy, transport and industry, and biodiversity.

Myanmar ratified the United Nations Convention to Combat Desertification (UNCCD) in January 1997 and even before that, the Forest Department had implemented a special greening project for the nine districts in 1994 (Forest Department 2009). The project area was extended to 13 districts with the formation of the New Dry Zone Greening Department in 1997. Rehabilitation and reforestation of mangroves is also being carried out in the Ayeyarwady delta region where large areas of mangrove forests had disappeared due to over-exploitation.

The Myanmar Agenda 21 (1997) was formulated by the National Commission for Environmental Affairs. The Agenda 21 facilitates integration of environmental considerations into national and sectoral development programs. The National Sustainable Development Strategy (2009) builds on the country's Agenda 21. Environmental considerations are taken into account in social and economic development to achieve sustainable development.

7.2.2 Vulnerable Sectors and Regions

Myanmar's agriculture, fisheries, and livestock sectors play a key role in supporting economic growth, local livelihoods, and food security. It is estimated that about 28.6% of Myanmar's GDP is from the agricultural sector (MoNREC 2017). The crop cultivation is mostly rain-fed and the majority of the farmers are small-scale landholders. The wide variety of agro-ecological tracts lead to farmers growing more than 60 different crops, including both tropical and temperate varieties. It is estimated that about 50% of Myanmar's agricultural land is cultivated with rice as the predominant food crop (Department of Meteorology and Hydrology 2012). Cereal crops include wheat, maize, and millet. Crops for domestic use are onions, potatoes, garlic, cassava, sweet potatoes, oil-seed (groundnut, sesame, sunflower), and pulses and beans (black gram, chickpea, soybean, pigeon pea). Cotton, sugarcane, jute, and rubber are cultivated as industrial crops for both domestic use and for export. Maize is grown predominantly for export and livestock feed. Fisheries in Myanmar are income for millions of people and a crucial source of livelihood. Fisheries contribute about 10% of the GDP and is a source of income for 5% of total population in Myanmar (MoNREC 2017). The marine fisheries and aquaculture are focused mainly along the lakes, rivers and the Ayeyarwady delta.

The backyard methods for livestock rearing contribute to most household income; and to a sizable portion of household capital in Myanmar. The livestock include cattle, buffalo, pigs, and poultry. At the national level, the country produces surplus food and exports some 1.8 million metric tons, contributing around 25% of export earnings, although geographic differences result in localized food shortages (MoNREC 2017).

According to NAPA (2012), Myanmar is affected by three main environmental stresses: (i) Climate related hazards and extreme weather events; (ii) Deforestation; and (iii) Diminishing water resources. Anthropogenic activities are identified as the root cause for the stresses. Myanmar experiences cyclones, strong winds, floods, storm surges, intense rains, extreme high temperatures, and droughts. Droughts are considered the most severe weather event in the country due to the effects on local communities, which include health impacts, damage to property and assets, as well as loss of income and livelihoods. Forests in Myanmar, especially in the lowland forests of the Tanintharyi Region, have been affected by degradation, shifting cultivation, and conversion to commercial oil palm plantations. The deforestation rate in Myanmar has been estimated at 466,420 hectares annually over the period 1989–1998. Furthermore, over the past two decades, Myanmar has lost more than 3% of its forest ecosystems (MoE 2009). The effects of land degradation, overutilization, and unfavorable agricultural practices have led to diminishing water supplies, especially for domestic, agricultural, and industrial usage. The shallowness of Inle Lake is a classic example of how degradation affected the largest lake in Myanmar. This has reduced the quantity of water available for use by local communities and threatened agricultural practices (floating gardens, fishing activities), tourism and recreational activities (Department of Meteorology and Hydrology 2012). An investigation of the linkages between hydrology, metal mining, heavy metal loading, and climate change in the Chindwin River projects that streamflow will decrease in the river under future climate conditions (Shrestha et al. 2019). Thus, it is critical to monitor the mines along this river as they discharge a large amount of toxic heavy metal loadings, contributing to reduced water quality and quantity.

7.3 Adaptation Priorities

7.3.1 Approach

A consultation workshop between Myanmar Climate Change Watch, SEEDS Asia, SEADPRI, ASEAN and ASEAN-India Green Fund was organized in 2015 to assess the adaptation needs and priorities under the national communication, and to identify the pilot activities, areas, and stakeholders from the relevant sectors at local levels. A total of 25 participants from government departments, local NGOs, and international NGOs, working for climate change activities in Myanmar participated in the workshop. Key players were Dr. Tun Lwin from Myanmar Climate Change Watch and Ms. Mitsuko Shikada from SEEDS Asia.

The participants were divided into three groups according to their interests and backgrounds for the group discussion. The first group was on Education, Health and Food Security, comprising members from Myanmar Climate Change Watch, Food Security Working Group, Green Lotus NGO, American Red Cross and SEEDS Asia. The second group was on Environment, Forest and Agriculture with the members from Myanmar Green Network, Renewable Energy Association Myanmar, Myanmar-Agro Action, BENCA, FREDA, Myanmar Forest Association, SEEDS Asia, and World Vision Myanmar. The third group was on Early Warning, Water, Energy & Transport with members from the Department of Meteorology & Hydrology, Plan International, Green Lotus, Renewable Energy Association Myanmar, Myanmar Red Cross Society, and SEEDS Asia. This resulted in four proposals being identified:

- Developing climate change adaptation-related education materials for children and people targeting different ecosystems such as Dry Zone, Coastal Zone, and Hilly Zone regions;
- Risk awareness and training of farmer groups in Dry Zone;
- Watershed management in Hlaw Gar Watershed Area, Yangon Region;
- Climate Change Adaptation Practices in Dry Zone (Integrated Resource Planning) focusing on Pwint Phyu Township, Magway Region in Dry Zone.

7.3.2 Potential Local Level Pilots

The potential pilots are identified based on the vulnerability to climate change, and socioeconomic sectors and communities including geographic areas, vulnerable groups, and livelihood types. The pilot areas were selected to represent the most vulnerable regions in Myanmar such as the dry zone and coastal areas.

Education and Food Security: Myanmar has different characteristic ecosystems which need different adaptation processes and knowledge. Thus, it is proposed that education materials be prepared targeting different ecosystems on climate change adaptation with specific focus on food security issues. The Dry Zone in the central part of Myanmar covers large parts of Magway, Mandalay, and lower Sagaing Regions, and it is one of the most food-insecure areas of the country (Fig. 7.1). Water shortage is the most crucial issue in that area due to the irregular and scarce rainfall. According to the food security assessment in the Dry Zone area conducted by World Food Program in 2011, the food security situation in that area has improved compared with 2009 and 2010. The report indicated that 17% of households are severely food insecure, 24% moderately. and only 59% are food secure. The mountain areas have their own specific issues on forest management and forest-related agriculture products. The coastal area, which is the rice bowl of the country, faces different issues related to climate change. The knowledge and

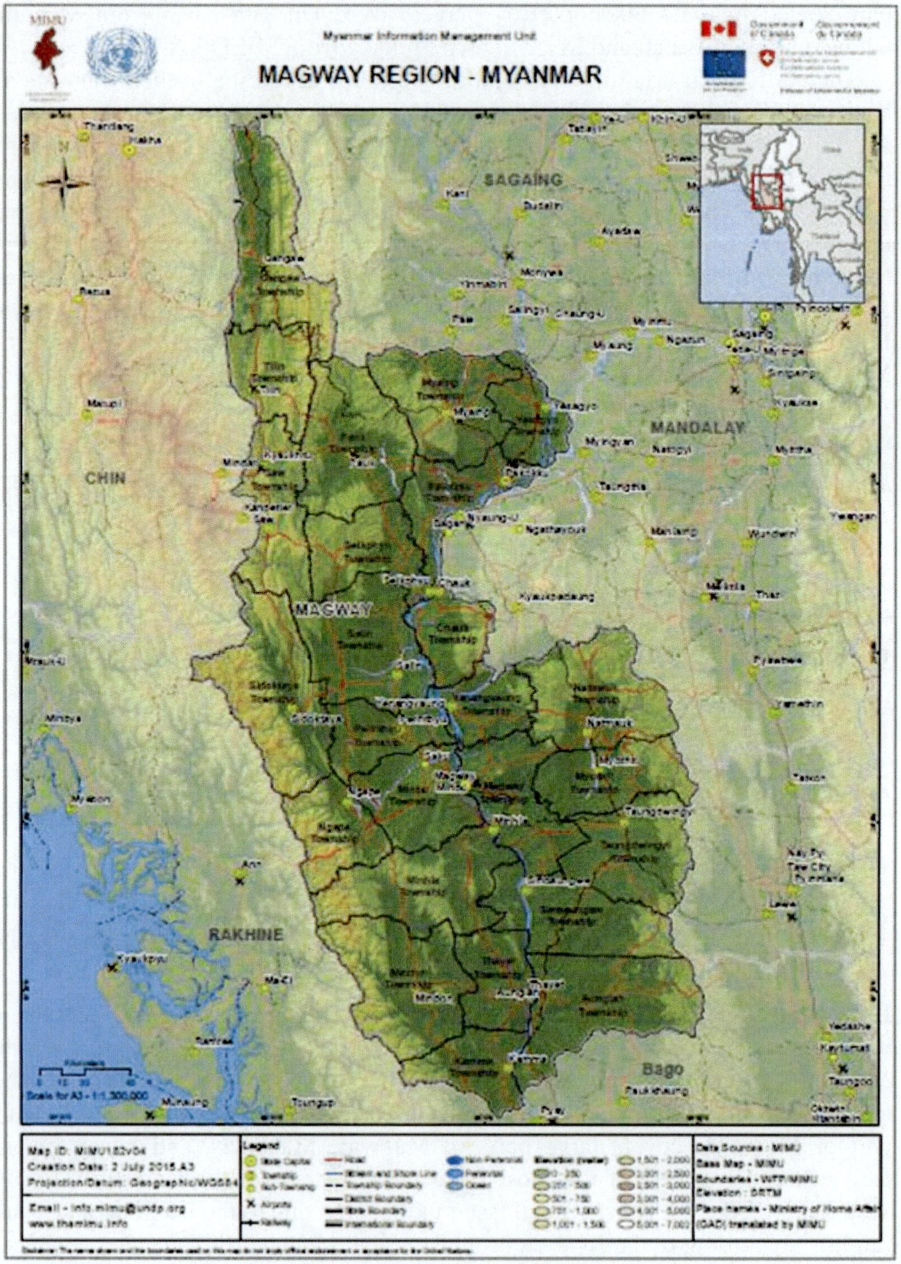

Fig. 7.1 Location map of the proposed pilot in Magway, Myanmar

information on adaptation measures for food security due to climate change is limited in the local levels and the local communities are not aware of policy and institution mechanisms. Although Village Disaster Management Committees have been formed in each village community, the institutional mechanism is not functioning well. In addition, coordination among the stakeholders such as local government, community-based organizations (CBOs), and local community is poor and information accessibility is limited among the community level. It was found that there is not enough user-friendly education material for climate change adaptation and global warming at the community level, and there are some language barriers in ethnic areas like Chin, Shan States, etc. Moreover, the training and learning materials are also very limited.

There are several stakeholders in that area that could implement the project, such as General Administration Department, Relief and Resettlement Department, Department of Meteorology and Hydrology, Ministry of Agriculture and Irrigation, Township Development Committee, CBOs, NGOs, private sectors, universities like the Agricultural University, media and also village-level volunteers from the Myanmar Red Cross Society. Furthermore, with the Building Resilience and Adaptation to Climate Extremes and Disasters (BRACED) project as a national platform, the Food Security Working Group which consists of CBOs, NGO networks, DRR Working Group could also be the network among national and local levels for the implementation of the project. Expert technical input from Indonesia, the Philippines, Vietnam, and the Regional Integrated Multi-Hazard Early Warning System for Africa and Asia (RIMES) could support the project.

Risk Awareness and Training: In 2011, Agriculture contributed ~43% to GDP and making up to 11% of export earnings. Overall trends of climate change in Myanmar indicate that the temperature has increased in the last six decades (~0.8 °C per decade) in the whole country. Moreover, the frequency of floods and droughts has also been increasing and resulted in the negative impacts on the agriculture sector, especially crop production. The Dry Zone is chronically receiving less rainfall compared with other parts of Myanmar, and as a result, it has a poor vegetation index (Fig. 7.2). Particularly, local rice varieties are no longer adapted to local climate conditions. Due to the extreme climate condition, traditional drying, harvesting, and storing practices are not effective in agriculture production. Thus, the local farmers need to have knowledge on climate change such as temperature and extreme weather events, and the adaptation measures in order to ensure resilience of harvesting, drying, and storing methods for future flood and drought events. In addition, conserving plant genetic resources, breeding new and improved varieties, land management, and capacity building is required to ensure food security under the future climate change impacts.

Again, this project can be implemented by stakeholders like General Administration Department, Relief and Resettlement Department, the Department of Meteorology and Hydrology, the Ministry of Agriculture and Irrigation, Township Development Committee, CBOs, NGOs, the private sector, the Agricultural University, media, local farmers, and Myanmar Rick Crop Entrepreneur Agency. And with the BRACED project as a national platform, the

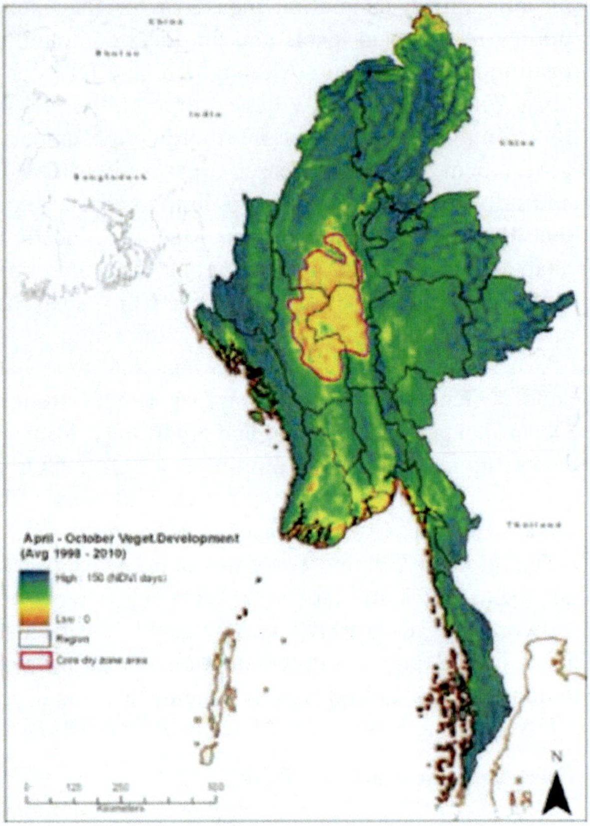

Fig. 7.2 Average vegetation index, 1998–2010. *Source* WFP (2011)

Food Security Working Group that consists of CBOs, NGOs network, DRR Working Group can also be the network among national and local level for the implementation of the project. Some technical input from experts from Indonesia, the Philippines, Vietnam and RIMES can also support the project.

Watershed Management: Hlawga Wildlife Park was established in 1982 by the Forest Department in Yangon Region with the objectives of providing environmental education facilities, protecting the forest and plants in the catchment area of the Hlawga Lake, and establishing a collection of indigenous plants and wildlife species (Fig. 7.3). In the year 2010, the site was jointly managed by the National Wildlife Conservation Division and private entities. The site preserves three types of habitats: evergreen forests, mixed deciduous forests, and swamp forests. There are 108 tree species and 12 mammal species, and 191 resident and migratory birds in the park. The rangers have been assigned and trained by the Forest Department. In 2010, due to a joint management with a private company, the staff were reorganized and infrastructures were renovated, but unfortunately, it was not successful. A major challenge for community forestry includes limited community participation and weak institutional systems for monitoring and conflict

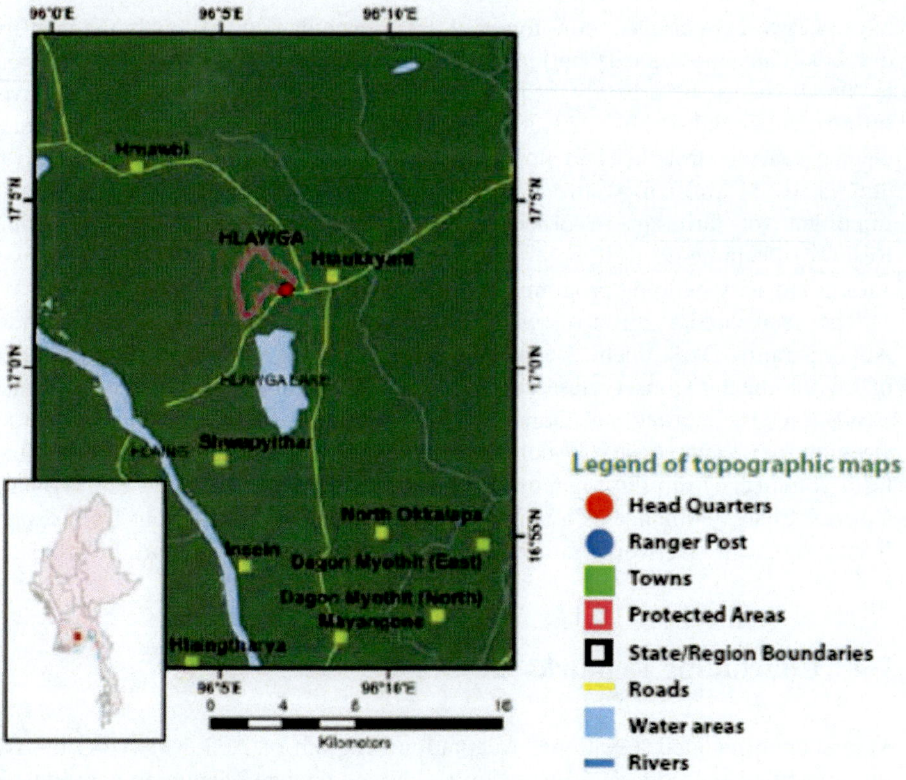

Fig. 7.3 Location map of the Hlawga watershed, Myanmar

resolution (Lin et al. 2019). Firewood exploitation and deforestation are the major issues in that area. Thus, the awareness program on the watershed management for the local community and staff from the Forest Department is required in that area (Forest Department 2009). Moreover, an education program on organic farming to reduce the water pollution, to get the behavior of reducing the use of chemicals in farming for the sustainable agriculture program is needed. Additionally, capacity building training program for fuel-efficient stove-making and provision of fuel stoves can be one of the adaptation measures for deforestation.

The stakeholders that could help implement the project in that area: General Administration Department, Ministry of Agriculture and Irrigation, Forest Department, CBOs, NGOs, and private sectors, universities such as the Agricultural University. Technical input from experts from International Nature Farming Research Centers and Asia Pacific National Agriculture Network can help support the project.

Early Warning: Recently, severe floods occurred in the central Dry Zone in July and August because of heavy rainfall; it caused huge amounts of loss and damage in not only lives but also properties. Climate change scenarios predict that

Myanmar will experience more frequent occurrence of flood events in the next five decades. Thus, the capacity building for the community to prepare for and respond to flood events is urgently required to ensure the safety of people's lives, infrastructure and livelihoods. Moreover, the labour migration due to climate change causing droughts and floods is the critical problem; it is expected to be around 10–11 million. Cattle are the most vulnerable to floods and they are important for farming. In order to take adaptation measures, the Integrated Resource Planning needs to be implemented among the different stakeholders with various capacity building programs, workshops, and meetings.

The stakeholder groups who could implement the project: General Administration Department, Department of Meteorology and Hydrology, Ministry of Environmental Conservation and Forestry, Township Development Committee, CBOs, NGOs, International Organizations from DRR Working Group which are working for Climate Change Adaptation in Myanmar, the private sector and village level volunteers from the Myanmar Red Cross Society. In addition, the Myanmar Climate Change Alliance as a national platform, and technical input from experts from Asia Pacific Network, Greater Mekong Region can also support the project.

7.4 Concluding Remarks

Myanmar formulated a National Adaptation Program of Action (NAPA) in 2012 that identifies and communicates priority climate change adaptation activities for the country. The three main challenges in the country are climate related hazards and extreme weather events, deforestation, and diminishing of water resources. Drought is the most severe weather event in the country affecting local communities, causing health impacts, damage to property and assets as well as loss of income and livelihoods. Potential local level pilots have been identified based on socioeconomic aspects, geographic settings, and vulnerability to climate change. The pilot areas represent the most vulnerable regions in Myanmar, comprising the dry zone and coastal region.

Acknowledgements The authors acknowledge the ASEAN-India Green Fund for supporting the ASEAN-India Project on "Enhancing Local Level Climate Change Adaptation in Southeast Asia" (Phase 1). This chapter draws partly on findings of the ASEAN-India Project, specifically the local level needs assessment conducted in 2015. The support of Ms. Han Swe from the Department of Meteorology and Hydrology Myanmar, Dr. Tun Lwin and Prof. Khin Maung Cho from Myanmar Climate Change Watch, and Miss Thinn Hlaing Oo from SEED Asia (Myanmar Project Office) is much appreciated.

References

Department of Meteorology and Hydrology (2012) The Myanmar national adaptation programme of action (NAPA) to climate change 2012. Ministry of transport republic of the Union Myanmar. http://unfccc.int/resource/docs/napa/mmr01.pdf. Accessed 21 Jan 2019

Forest Department (2009). Forest department annual report. Ministry of environmental conservation and forestry of Myanmar

Lin T, Htun KT, Gritten D, Martin AR (2019) The contribution of community forestry to climate change adaptive capacity in tropical dry forests: lessons from Myanmar. Int For Rev 21(3):324–340(17)

MoE W (2009) Burma taking severe hit from climate change: watchdog. http://reliefweb.int/node/336464. Accessed 21 Jan 2019.

MoNREC (2017) Myanmar climate change strategy and action plan. The Republic of the Union of Myanmar, Nay Pyi Taw, Myanmar. https://policy.asiapacificenergy.org/sites/default/files/MCCSAP-Feb-Version.pdf. Accessed 1 Jan 2019

National Commision for Environmental Affairs, Ministry of Forestry National and United Nation Environment Programme, Regional Center for Asia and the Pacific, Sustainable Development Strategy for Myanmar (2009)

Shrestha S, Gunawardana SK, Piman T, Babel MK (2019) Assessment of the impact of climate change and mining activities on streamflow and selected metal loading in the Chindwin river. Myanmar Environ Res. https://doi.org/10.1016/j.envres.2019.108942

South A (2020) Myanmar's "Rohingya" conflict: book review. J Contempo Asia 50(1):168–170

Union of Myanmar (2009) Hazard profile of Myanmar. Naypyitaw, Myanmar. https://www.preventionweb.net/files/14567_14567HazardReport25.8.091.pdf

WFP (World Food Program) (2011). Food security assessment in the dry zone Myanmar. https://documents.wfp.org/stellent/groups/public/documents/ena/wfp234780.pdf?iframe. Accessed 21 Jan 2019

Chapter 8
Climate Change Adaptation in the Philippines

Juan M. Pulhin and Maricel A. Tapia

Abstract Climate change adaptation in the Philippines has evolved from "no regret" solutions to serious mainstreaming into policies and development plans, which specifically address the country's and the local governments' vulnerabilities. Several priorities identified in the National Climate Change Action Plan (NCCAP) through a gap analysis following an inventory of climate change researches have already been done in the country, and areas that need further studies, as well as new and emerging issues, are highlighted in this chapter. The most number of research conducted in the Philippines under the NCCAP is in the areas of environmental sustainability and food security. The proposed pilot is to take San Vicente, a municipality located in the northwestern side of the Palawan Province, as a single socioecological system. A holistic adaptation approach for building resilience to climate change will be taken, bearing in mind the concepts of multifunctionality, transdisciplinarity, participation, complexity, and sustainability, specifically targeting the sub-components of resilience.

Keywords Forest degradation · Kanan watershed · Climate change · Food security · Philippines

8.1 Introduction

Adaptation has gained significant attention in recent years with the varying impacts of changing climate, including variability and extreme weather events, which were experienced in different parts of the world. The demands for adaptation have become immensely apparent, especially with the current and predicted environmental risks. Utter disregard to the expected system transformation required for handling the impacts from environmental risks may render some areas and

J. M. Pulhin (✉) · M. A. Tapia
University of the Philippines, Los Baños, Philippines
e-mail: jmpulhin@up.edu.ph

M. A. Tapia
e-mail: matapia@up.edu.ph

J. J. Pereira et al. (eds.), *Climate Change Adaptation in Southeast Asia*, Disaster Risk Reduction, https://doi.org/10.1007/978-981-16-6088-7_8

economic activities unviable (Nelson et al. 2007). Further, climate change is a multiplier of existing disadvantages; hence doing nothing would also lead to unprecedented losses much greater than the costs of adaptation, as emphasized by the Stern Report (2007).

The Philippines, being a developing country and an archipelagic nation that sits adjacent to the Pacific Ocean, is highly exposed and sensitive to various climatic threats, which are expected to increase in intensity and frequency with the changing climate. These threats include more intense tropical cyclones, drastic changes in rainfall patterns, sea-level rise, and increasing temperatures. The Global Climate Risk Index 2015 ranked the Philippines as number one in the list of the countries most affected by the impacts of weather-related loss events (storms, floods, heat waves, etc.) in 2013. This is also the same year when Typhoon Haiyan hit the country, the strongest tropical cyclone on record to hit land, which caused over USD 13 billion in economic losses and 6,000 deaths. In the period 1994–2013, the Philippines ranked fifth among the most affected by climate disasters, with Honduras, Myanmar, Haiti, and Nicaragua topping the list. The rankings underscored that less-developed countries are more vulnerable to climatic risks compared to industrialized countries (Kreft et al. 2014).

Managing climate risks requires adaptation. Adaptation is defined as "the process of adjustment to actual or expected climate and its effects. In human systems, adaptation seeks to moderate or avoid harm or exploit beneficial opportunities. In some natural systems, human intervention may facilitate adjustment to expected climate and its effects" (IPCC 2014). Meanwhile, the legal documents in the Philippines (such as Climate Change Act or Republic Act (RA) 9729 and the National Framework Strategy on Climate Change or NFSCC) define adaptation as "the adjustment in natural or human systems in response to actual or expected climatic stimuli or their effects, which moderates harm or exploits beneficial opportunities". The latter was based on the IPCC Third (TAR) and Fourth (4AR) Assessment Reports, which were available when the Climate Change Act was passed in 2009 and the NFSCC was formulated in 2010 (IPCC 2001, 2007a,b).

Adaptation experience across the globe has grown and it is increasingly being embedded in planning processes (IPCC 2014). Adaptation is executed at various scales and levels, and its implementation differs depending on the context, such as resources, values, and needs. Current approaches to climate change adaptation (CCA), however, focus predominantly on sectoral and technological approaches. Apart from this, there is little discussion in the literature on how adaptation is realized in practice as well as the effects of adaptation strategies implemented (O'Brien and Hochachka 2010; IPCC 2014).

This chapter reviews the national documents on adaptation in the Philippines, including the local situation and priorities. It presents a historical background of the country's adaptive responses to climate change in terms of policies, programs and projects, and other strategies. It also highlights the framework for adaptation employed in planning and implementing strategies, as well as the stakeholders involved. Gaps and priorities in adaptation are presented based on existing literature.

8.2 National Perspective

8.2.1 National Arrangements and Institutions

While CCA appears to just receive significant attention only recently, the Philippines already has a long history of adapting to climate variability and extremes, being highly exposed to these hazards (World Bank 2010; Lasco et al. 2008a). For instance, farmers have long been planting drought- and pest-resistant crop varieties, and have been practicing crop rotation to manage soil fertility and pest infestation (World Bank 2010). The indigenous knowledge and practices of some Filipino communities have also helped them maintain a sustainable relationship with the natural environment, which is the source of their livelihood (Sandoval and Baas 2013). More recently, the concept and feasibility of managed retreat as an adaptation measure is also being explored (Doberstein et al. 2020).

Early strategies on climate change at the national level, however, were focused on disaster response. In 1978, the then President Ferdinand Marcos passed the Presidential Decree No. 1566, which strengthens the Philippine disaster control capability and establishes the national program on community disaster preparedness. Under this law, the National Disaster Coordinating Council was created under the Office of Civil Defense, which coordinates the activities and functions on disaster management, among others (http://www.gov.ph/1978/06/11/presidential-decree-no-1566-s-1978/).

The 1987 Constitution declared it the policy of the state to "protect and advance the right of the people to a balanced and healthful ecology in accord with the rhythm and harmony of nature" (Article II, Sect. 16). Consistent with this state policy, the Philippines, in 1989, took on sustainable development as a guiding principle in its development efforts through the approval and adoption of the framework of the Philippine Strategy for Sustainable Development (PSSD). The country's participation in the Rio Summit further developed these initiatives, which led to the establishment of the Philippine Council for Sustainable Development (PCSD) in 1992 and the formulation of the Philippine Agenda (PA) 21 in 1996 (Republic of the Philippines 2014; World Bank 2010).

The Philippines has been noted to possess "one of the most responsive institutional and legislative mechanisms for environmental management in Southeast Asia" (Tan 2013). It is therefore not surprising that even prior to its signing of the United Nations Framework Convention on Climate Change (UNFCCC) at the Rio Summit in 1992, the Philippines already took the step to respond to the impending problem of climate change. In 1991, the then President Corazon Aquino signed Administrative Order (AO) No. 220, creating the Inter-Agency Committee on Climate Change (IACCC). The IACCC was composed of various government agencies and an umbrella non-government organization, with the primary function of formulating policies and response strategies related to climate change which also shape the country's position in international negotiations (http://www.gov.ph/1991/05/08/).

The Philippines ratified the UNFCCC on August 2, 1994, and entered it into force on October 31, 1994. Under the Convention, countries agree to protect the climate system by stabilizing the greenhouse gas (GHG) concentrations in the atmosphere, on the basis of equity and in accordance with their common but differentiated responsibilities and respective capabilities. It is also a platform where governments share and gather information on GHG emissions and cooperate with regards to adaptation efforts (Republic of the Philippines 2014; World Bank 2010).

The Philippines signed the Kyoto Protocol to the UNFCCC in 1998 and ratified it in 2003. The Protocol is an international treaty that committed developed countries to emission reduction targets. The ratification of the Kyoto Protocol enabled it to participate in the Clean Development Mechanism (CDM), which is an international offset program where developing countries can reach sustainable development goals and increase investment flows through emission reduction credits that can be traded with industrialized countries (Republic of the Philippines 2014; World Bank 2010).

The CDM and other flexible mechanisms to meet the emission targets of industrialized countries gave rise to the carbon market. This development has pushed to the fore climate change mitigation as an identified priority in the Medium-Term Development Plan of the Philippines for 2004–2010 (World Bank 2010). Lasco et al. (2008b) also observed that climate change was only mentioned once in this document, and this was in the context of CDM under energy independence. Climate change adaptation was taken in the context of disaster risk reduction, which reflected the then prevailing approach. Sectoral plans that aimed to address climate change, such as the Philippine Energy Plan 2004–2013 of the Department of Energy (DOE), also focused on mitigation, which emphasized improving energy efficiency and promoting renewable energy sources (World Bank 2010). Aside from the above strategies, the following laws also consider climate change issues (World Bank 2010):

- Agriculture and Fisheries Modernization Act of 1997 (RA 8435), which directs the Department of Agriculture (DA), in coordination with the Philippine Atmospheric, Geophysical and Astronomical Service Administration (PAGASA) and other appropriate agencies, to devise a method for regularly monitoring and considering the effect of global climate changes, weather distribution, and annual productivity cycles for the purpose of forecasting and formulating agriculture and fisheries production programs.
- Philippine Clean Air Act of 1999 (RA 8749), which orders the Department of Environment and Natural Resources (DENR) to promote environmentally sound technologies for the handling of wastes with due concern on the effects of climate change and to prepare a national plan on the reduction of GHGs in the country consistent with the UNFCCC.
- Ecological Solid Waste Management Act of 2000 (RA 9003), which instructs the DENR, concerned agencies, and local government units (LGUs) to prepare and implement solid waste management plans.

- Philippine Clean Water Act of 2004 (RA 9275), which aims to reduce water pollution through better management of sewerage and sanitation, industrial effluents, and agricultural, industrial, and residential wastes.
- Biofuels Act of 2006 (RA 9367), which aims to reduce dependency on imported fuels and seeks to achieve energy self-sufficiency, while at the same time mitigating toxic and GHG emissions.
- Renewable Energy Act of 2008 (RA 9513), which promotes the development and utilization of renewable energy resources in the context of energy security and climate change imperatives.

The undisputed evidence of science that "the warming of the climate system is unequivocal", has brought a wave of urgency to respond to climate change (IPCC 2007a). In February 2007, the Presidential Task Force on Climate Change (PTFCC) was created with the issuance of AO No. 171. The DENR served as the chair of the task force, and its functions included conducting a rapid assessment on the impacts of climate change in the Philippines, ensuring strict compliance with air emission standards, and combating deforestation and environmental degradation. The IACCC, which was formed in 1991, acted as the technical arm of the PTFCC. Later in the same year, AO No. 171-A was passed transferring the chairmanship of the task force to DOE (Republic of the Philippines 2014). The Advisory Council on Climate Change Mitigation, Adaptation, and Communication was also created through DENR Special Order No. 2007-653, replacing the IACCC as the technical arm of the PTFCC (Finkel and Duhaylungsod 2010).

The PTFCC was reorganized in December 2008 through Executive Order (EO) No. 774, making the President of the Philippines as its chair and the Cabinet members as the members. Sixteen task groups were also formed, which included solid waste management, watershed protection, rainwater conservation, and fossil fuels, among others (http://www.gov.ph/2008/12/26/executive-order-no-774-s-2008/). Meanwhile, EO 785 issued in February 2009 mandated the PTFCC to develop a National Climate Change Framework that will consolidate and institutionalize government initiatives to achieve synergy in the implementation of climate change programs and projects. Concurrent with these initiatives, several bills on climate change were passed at the lower and upper chambers of the Congress, which would bring changes to the then existing task force and national strategies on climate change (World Bank 2010).

The aftermath of Typhoons Milenyo (Xangsane) and Reming (Durian) that hit the Philippines in 2006 has sparked a heightened level of consciousness on the necessity of CCA, especially at the local level. Being at the receiving end of frequent and severe climate hazards due to its geographical location, the Provincial Government of Albay took a proactive role in addressing disaster and climate risks and proclaimed CCA as its governing policy (PGA and CIRCA 2010; Uy et al. 2011). The Albay Province has initiated innovative approaches toward disaster risk reduction and CCA. Consistent with PD 1566, Albay is the first and only province in the Bicol region to establish the Provincial Disaster Management Office (PDMO). This office coordinates with various stakeholders to promote efficient intervention on disaster preparedness, emergency response, and recovery. PDMO was later institutionalized to the Albay Public Safety and Emergency Management Office (APSEMO) in 1994, an independent department that serves as the technical secretariat and administrative arm to support the Provincial Disaster Coordinating Council of Albay. This move has strengthened the disaster management capability of the province and ensured continuity of related programs (PGA and CIRCA 2010; Lasco et al. 2008b; Republic of the Philippines 2014).

In 2007, under the helm of the then newly elected governor, Jose Maria Clemente Salceda, the provincial government of Albay launched the Albay in Action on Climate Change (A2C2), a pioneering program for local climate change adaptation. It aims to embed disaster risk reduction to enforce climate-proofing and disaster-proofing of development. A2C2 spearheaded programs, such as a 90-hectare mangrove plantation project in Manito, Albay, and other land use and land use change and forestry (LULUCF) activities. It also led to the integration of the CCA program in the curricula of all schools, colleges, and universities in Albay (PGA and CIRCA 2010; Lasco et al. 2008b).

Aside from local actions, the province also brought the discourse on CCA to the national level by convening two National Conferences on Climate Change Adaptation (NCCCA). The first NCCCA held in Albay in October 2007 produced the Albay Declaration on Climate Change, which called for the passage of a policy prioritizing CCA in the national and local agenda. The Declaration has served as a model framework for integrating climate change concerns into national and local planning, leading to climate change adaptation. It was also instrumental in the passage of the Climate Change Act after two years. The second NCCCA held in October 2009 adopted the Manila Declaration, which made an appeal to the President of the Philippines for urgent action on the then pending bill on disaster reduction management (PGA and CIRCA 2010; Uy et al. 2011; Republic of the Philippines 2014).

The Province of Albay exemplifies the "living with risks" concept and proved that geography and environmental phenomena are not obstacles in the pursuit of development. With political will, forged partnerships, and legislative enactments, it has succeeded in rising above disasters and achieving a stronger economy (Espinas 2012). The good practices that it has pioneered has made the province a prototype for local CCA not only in the Philippines but also across the globe, with the United Nations International Strategy for Disaster Reduction (UNISDR) and the World Bank declaring Albay as a "Global Local Government Unit Model for Climate Change Adaptation" in 2008 (PCA and CIRCA 2010).

8.2.2 *Climate Change Adaptation at the National Level*

The Philippines made another landmark legislation with the passage of the Climate Change Act (RA 9729) in 2009. In a nutshell, RA 9729 created an enabling environment to bolster climate change governance in the country in compliance with international frameworks (such as the UNFCCC) and in line with national and local development initiatives (Adaptation Knowledge Platform 2012). It established the Climate Change Commission (CCC) as the sole policy-making body mandated to coordinate, monitor, and evaluate the programs and action plans related to climate change. CCC is housed under the Office of the President and is being chaired by the President. Three Commissioners, one of whom serves as the Vice-Chairperson, also sit on CCC, and it has an advisory board composed of different government agencies and sectoral representatives (Republic of the Philippines 2014).

RA 9729 highlights the need for appropriate adaptation in recognition of the country's vulnerability to climate change. The following adaptation-related concepts, actions, and interventions were stated in the Act, "adaptation planning; support for research and development programs; dissemination of information on CCA; enhancement of adaptive capacities of vulnerable human settlements; identification of options and prioritization of adaptation measures; and integration of climate change mitigation and adaptation framework" (Adaptation Knowledge Platform 2012).

In 2010, the National Framework Strategy on Climate Change (NFSCC) 2010–2022 was formulated, following the provision of the Climate Change Act. The Framework serves as the basis for climate change planning, research and development, and other related actions. It was also developed based on the needs of the country, particularly in relation to identified vulnerabilities, adaptation and mitigation potential (Climate Change Commission 2010).

NFSCC envisions a "climate risk-resilient Philippines with healthy, safe, prosperous and self-reliant communities, and thriving and productive ecosystems". Its goal is "to build the adaptive capacity of communities and increase the resilience of natural ecosystems to climate change, and optimize mitigation opportunities towards sustainable development". As illustrated in Fig. 8.1, increasing temperatures, changing rainfall patterns, sea-level rise, and extreme weather events are the climate change hazards that the country specifically contends with, which impede the achievement of sustainable development. The path toward sustainable development, therefore, requires tackling the impacts to and vulnerabilities of different sectors from the above hazards. At the same time, sustainable development affects the capacity to adapt to the impacts and address the vulnerabilities to climate change (Climate Change Commission 2010).

Adaptation, mitigation, and cross-cutting strategies (such as capacity development, knowledge management, etc.) would be the means for addressing climate change impacts and vulnerabilities. These shall be implemented through multi-stakeholder partnerships, financing, valuation, and policy, planning, and mainstreaming. Adaptation, however, is given high priority due to the Philippines' geophysical and socioeconomic characteristics. The Framework's adaptation thrusts include enhanced vulnerability and adaptation assessment, integrated ecosystem-based management, water governance and management, climate-responsive agriculture, climate-responsive health sector, climate proofing of infrastructures, and disaster risk reduction. It requires that the critical aspect of

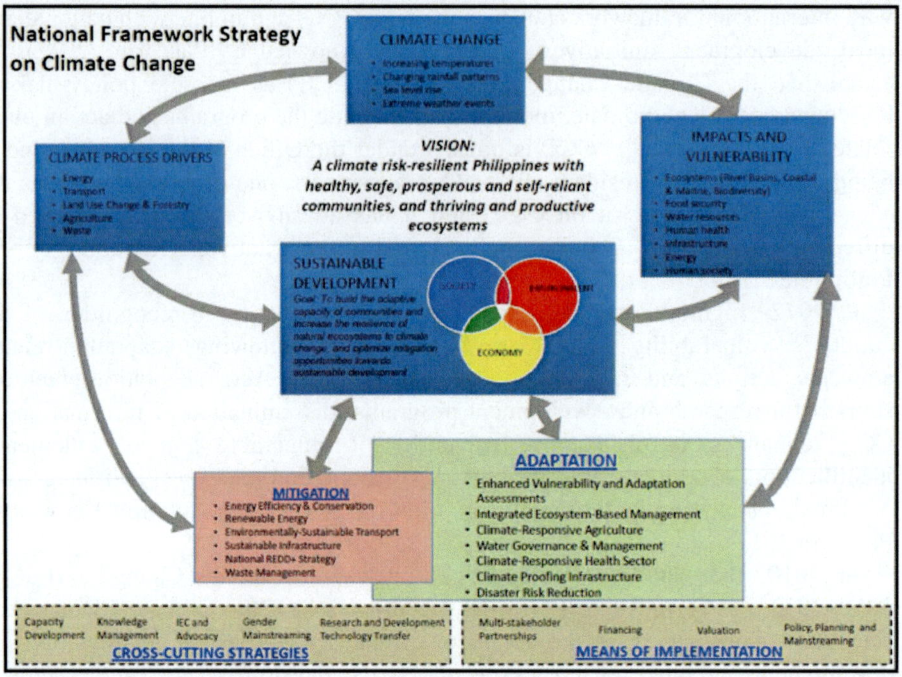

Fig. 8.1 National framework strategy on climate change

adaptation be implemented at all levels of governance. Mitigation strategies should also be implemented in the context of adaptation (Climate Change Commission 2010; Adaptation Knowledge Platform 2012).

To realistically achieve the country-driven programs on adaptation (including mitigation), the NFSCC was translated into the National Climate Change Action Plan (NCCAP) 2011–2028. Consistent with the NFSCC, the NCCAP aims "to build the adaptive capacities of women and men in their communities, increase the resilience of vulnerable sectors and natural ecosystems to climate change, and optimize mitigation opportunities towards gender-responsive and rights-based sustainable development" (Climate Change Commission 2011).

The NCCAP identified seven strategic priorities: food security, water sufficiency, environmental and ecological stability, human security, climate-friendly industries and services, sustainable energy, and knowledge and capacity development (Fig. 8.2). The achievement of these seven intermediate objectives would lead to enhanced adaptive capacity of communities, resilience of natural ecosystems, and sustainability of built environment to climate change. This desired result, together with the direct contribution from climate-smart industries and services, sustainable energy, and knowledge and capacity development, is seen as instrumental for achieving the ultimate outcome of a successful transition toward climate-smart development (Climate Change Commission 2011).

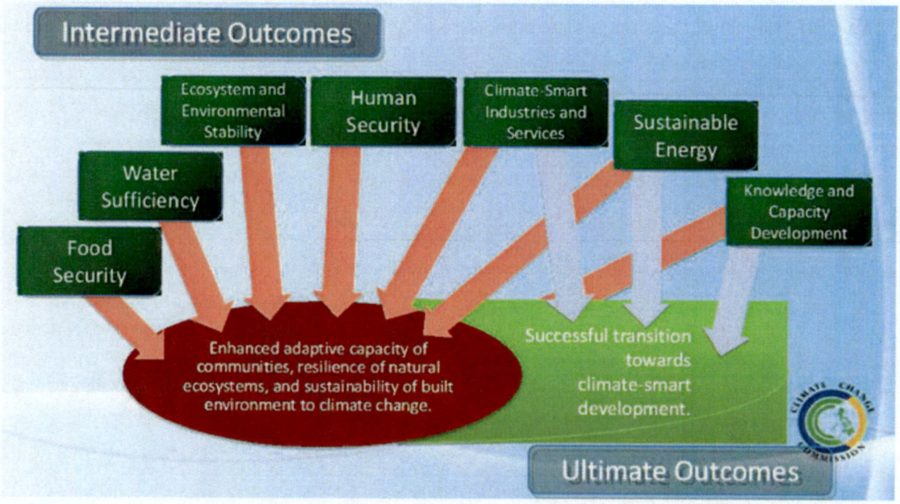

Fig. 8.2 Overview of the national climate change action plan. *Source* Sanchez (2015)

Each of the seven strategic priorities is expressed as an objective, with specified immediate outcomes (Table 8.1) and accompanying outputs and activities. It can be observed, however, that only five of the priority areas are directly contributing to the goal of CCA. These are food security, water efficiency, environmental and ecological stability, human security, and knowledge and capacity development. The other two are geared toward mitigation.

While the Climate Change Act also addresses disaster risks, it was found wanting in disaster risk reduction and management (DRRM). Hence, less than a year after the passage of the above law, and heeding the call made by various experts and stakeholders after the second NCCCA, the Philippine Disaster Risk Reduction and Management Act (RA 10121) was legislated.

The Philippine Disaster Risk Reduction and Management Act of 2010 defines DRRM as "the systematic process of using administrative directives, organizations, and operational skills and capacities to implement strategies, policies and improved coping capacities in order to lessen the adverse impacts of hazards and the possibility of disaster (Guingona and Bizaon 2009). Prospective disaster risk reduction and management refers to risk reduction and management activities that address and seek to avoid the development of new or increased disaster risks, especially if risk reduction policies are not put in place". Disaster risk reduction (DRR), on the other hand, is referred to in the same law as "the concept and practice of reducing disaster risks through systematic efforts to analyze and manage the causal factors of disasters, including through reduced exposure to hazards, lessened vulnerability of people and property, wise management of land and the environment, and improved preparedness for adverse events".

One of the welcome provisions of the new law is the availability of 70% of the Local and National Disaster Risk Reduction Management Fund (LDRRMF and

Table 8.1 Objectives and immediate outcomes of the seven NCCAP priority areas

Priority area	Objective	Immediate outcomes
Food security	Ensured food availability, stability, access, and safety amidst increasing climate change and disaster risks	1. Enhanced climate change resilience of agriculture and fisheries production and distribution systems 2. Enhanced resilience of agriculture and fishing communities on climate change
Water security	Water resources are sustainably managed and equitable access is ensured	1. Water governance restructured toward a climate and gender-responsive water sector 2. Sustainability of water supply and access to safe and affordable water ensured 3. Knowledge and capacity for CCA in the water sector enhanced
Ecosystem and environmental sustainability	Enhanced resilience and stability of natural systems and communities	1. Ecosystems protected, rehabilitated, and ecological services restored
Human security	Reduced risks of men and women and other vulnerable groups (children, elderly and persons with disability, etc.) from climate and disasters	1. CCA and disaster risk reduction practiced by communities and sectors at all local levels 2. Health and social protection delivery systems are responsive to climate change risks 3. Climate change-adapted human settlements and services developed, promoted, and adopted
Climate-smart industries and services	Climate change-resilient, eco-efficient, and environment-friendly industries and services developed, promoted, and sustained	1. Climate-smart industries and services promoted, developed, and sustained 2. Sustainable livelihood and jobs created from climate-smart industries and services 3. Green cities and municipalities developed, promoted, and sustained
Sustainable energy	Sustainable and renewable energy and ecologically efficient technologies adopted as major components of sustainable development	1. Nationwide energy efficiency and conservation program promoted and implemented 2. Sustainable and renewable energy (SRE) development enhanced 3. Environmentally sustainable transport promoted and adopted 4. Energy systems and infrastructures climate-proofed, rehabilitated, and improved
Knowledge and capacity development	Enhanced knowledge and capacity of women and men to address climate change	1. Enhanced knowledge on the science of climate change 2. Capacity for CCA, mitigation and disaster risk reduction at the local and community level enhanced 3. Climate change knowledge management established and accessible to all sectors at the national and local levels

NDRRMF, previously called calamity fund) for disaster mitigation and preparedness, while 30% will be allocated for quick response activities. This was a decisive shift from the earlier practice of disbursing the calamity fund only after the occurrence of a disaster, which provides little or no financial resources for pre-emptive response (Republic of the Philippines 2014).

Mitigation, as a strategy in DRRM, is defined in RA 10121 as "structural and non-structural measures undertaken to limit the adverse impact of natural hazards, environmental degradation, and technological hazards and to ensure the ability of the at-risk communities to address vulnerabilities aimed at minimizing the impact of disasters. Such measures include, but are not limited to, hazard-resistant construction and engineering works, the formulation and implementation of plans, programs, projects and activities, awareness raising, knowledge management, policies on land-use and resource management, as well as the enforcement of comprehensive land-use planning, building and safety standards, and legislation".

The above disaster response overlaps with CCA, particularly in the area of climate-proofing development initiatives. DRRM and CCA address different but related problems, with the former focused on an ongoing problem (disasters) and the latter concerned with an emerging issue (climate change). Their convergence lies in the common factor of weather and climate, and the similar tools to monitor, analyze, and address the impacts (UNISDR 2009). As illustrated in Fig. 8.3, the climate- and weather-related hazards that usually cause disasters are also the same extreme weather events that are predicted to be more frequent and intense with the changing climate. The gradual increase in climatic parameters, such as sea-level rise, rising sea temperature, and changes in precipitation patterns, may also affect

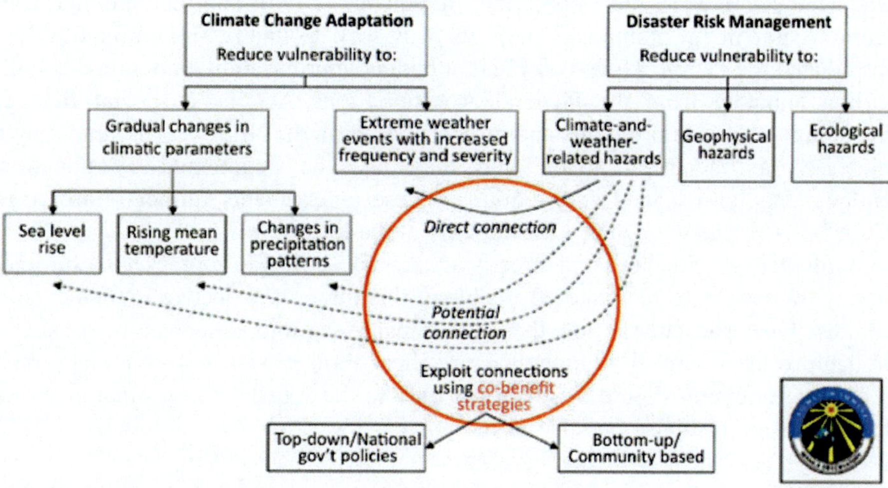

Fig. 8.3 Conceptual linkages of climate change adaptation and disaster risk management. *Source* Climate Change Commission (2011)

the occurrence of meteorological hazards. With this, it is therefore logical and beneficial to have synergistic measures for CCA and DRRM.

Disaster risk reduction (DRR) is emphasized in the NCCAP through the priority areas on human security and knowledge and capacity development. Under these intermediate outcomes, enhanced knowledge on CCA and DRRM and their integration in local plans are among the expected outputs (Climate Change Commission 2011). RA 10121, on the other hand, further strengthened CCA, particularly in view of the DRRM-CCA linkage (Adaptation Knowledge Platform 2012). The palpable linkages between CCA and DRR have created a legislative impetus toward their integration. Various challenges, however, hinder the full realization of this initiative. These include mismatches in scale, norms, and knowledge (De Leon and Pittock 2016). Scale mismatches could involve institutional, temporal, vertical (level), and horizontal (sectoral) disparities. For instance, CCA deals with incremental long-term changes at the global scale while DRR on instantaneous events embodied by disasters. Thus, dominant actions focused on the latter rather than the sources of vulnerability. Current norms on responding to climate-related disasters also rely on formal approaches, involving the development of strategic frameworks and national and local plans. Such responses may be problematic for climate change as the latter involves a higher level of uncertainty. Knowledge mismatch is seen in the highly technical and largely upper scale (global, regional, and national) generation of climate change assessments, which somehow restricts its usage by disaster practitioners and related stakeholders.

Meanwhile, the Climate Change Act was amended in 2012 through the People's Survival Fund Act (RA 10174). It included the Department of Budget and Management (DBM), the Department of Finance (DOF), and the National Youth Commission in the composition of its Advisory Council. More importantly, it introduced the People's Survival Fund, a budget appropriation amounting to PhP 1 billion annually from the General Appropriations Act, as a special fund for financing adaptation programs and projects based on the NFSCC. The fund may be increased as the need arises, and may also be augmented by donations, endowments, grants, and contributions. These amendments further reinforce the CCA initiatives in the country, particularly at the local level.

With three specific laws that directly address CCA, the Philippines has a sufficient policy framework to facilitate efforts along this line. The effective implementation of these laws, particularly how these are transformed into actions at the local level to achieve the vision of a climate-risk resilient Philippines, is what remains to be seen. Due attention should be given not only to the specific actions that constitute the adaptation response but also to the process by which these are realized (IPCC 2014).

8.2.3 Climate Change Adaptation at the Local Level

The implementation of NCCAP at the local level is demonstrated following the concept of ecologically stable and economically resilient town or eco-town. An eco-town is a planning unit composed of municipalities or a group of municipalities located within and around boundaries of critical key biodiversity areas, which are at high risk to climate change. It will be built around protected areas and key biodiversity areas, using ecosystem management-based approach to enable communities to become ecologically and economically stable (CCC and GGGI 2014).

While the concept of eco-town is not unique to the Philippines, the country's application of this framework focuses on adaptation. It presents an innovative approach to integrate and mainstream CCA and DRR in local development plans, programs, and activities to promote green growth. Through this, the eco-town project also highlights the role of the local government units (LGUs) as frontliners in responding to the risks of climate change. The eco-town demonstration sites include San Vicente, Palawan; Siargao Island, Siargao; Eastern Samar; Batanes; and Upper Marikina River Basin and Upper Landscape (CCC and GGGI 2014).

The Philippines' eco-town framework adopts the ecosystem-based approach (EBA) in implementing adaptation. EBA, as defined by the Convention on Biological Diversity, is "a strategy for the integrated management of land, water and living resources that promotes conservation and sustainable use in an equitable way". It follows a ridge to reef approach in maintaining the natural structure and functions of ecosystems (CCC and GGGI 2014). The process of implementing the eco-town framework requires the use of different assessment tools to come up with comprehensive adaptation measures (Fig. 8.4). The framework also consists of seven components (Climate Change Commission 2010): (i) Natural resource

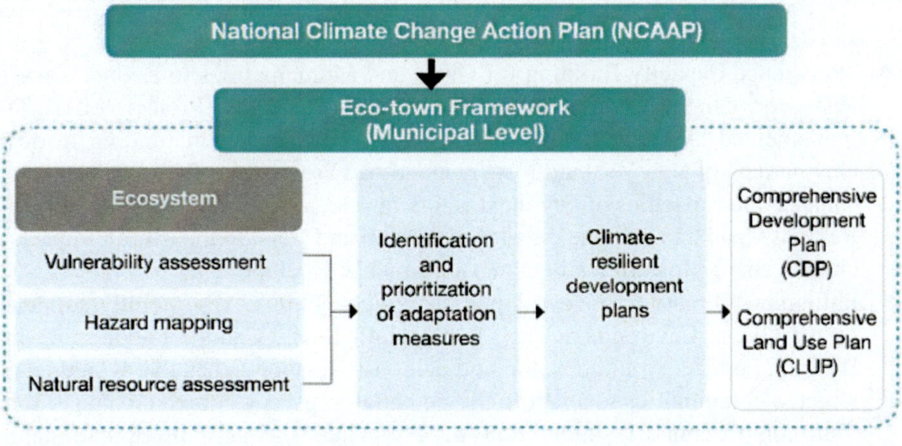

Fig. 8.4 The Eco-town framework for implementation of the NCCAP at the local level. *Source* CCC and GGGI (2014)

assessment; (ii) Vulnerability assessment; (iii) Environment and natural resources accounting; (iv) Initiation of adaptation and mitigation measures; (v) Application of climate adaptation support service; (vi) Design and implementation of financing scheme(s); and (vii) Integration of climate change into local plans.

One of the successfully completed eco-town projects was the case of San Vicente, Palawan. Following the above process, a menu of adaptation measures was developed for different sectors. For instance, modernization of farming practices and the introduction of technology to increase productivity characterized the preferred adaptation strategies for agriculture. Prioritized measures for the coastal and marine sector focused on protecting the coral reefs, sustainable utilization of marine resources, and disaster preparedness. Those identified for the health sector include provision of clean and adequate water supply system. It is yet to be seen if San Vicente, Palawan will live up to the label of an eco-town through the successful implementation of the above CCA (CCC and GGGI 2014).

Aside from the eco-town project, the following are past and ongoing programs and projects on CCA in the Philippines (Adaptation Knowledge Platform 2012):

- Project Climate Twin Phoenix—implemented by the CCC, and with technical assistance from the United Nations Development Programme (UNDP) and Australian Agency for International Development (AusAID). This project is envisioned to be the first stage of a more comprehensive and long-term capacity development program for cities and municipalities to manage risks from climate change and climate-related natural hazards. The project covers areas in Mindanao that were hardest hit by Typhoons Sendong (Washi) in 2011 and Pablo (Bopha) in 2012. Project activities include creation of flood hazard maps and preparation of Flood Contingency Plans.
- Supporting the CCC in the Implementation of the NCCAP—the German Agency for International Cooperation (GIZ) supported CCC and DOE to improve adaptation capacity and enhance mitigation. It sought to increase local level capacities for the development and implementation of climate strategies and action plans.
- "Resilience Capacity Building for Cities and Municipalities to Reduce Disaster Risks from Climate Change and Natural Hazards, Phase 1" or Project ReBUILD—implemented by CCC, in partnership with UNDP and with funding from the Government of New Zealand/New Zealand Aid Programme. This project aims to enhance the capacities of key local actors in selected cities and municipalities in Regions 2 and 6 to manage disaster risks from and adapt to the overall impacts of climate change toward resilient and sustainable development.
- Philippine Climate Change Adaptation Project (PhilCCAP)—jointly supported by the Global Environment Facility, Special Climate Change Fund, and World Bank, the project would develop and demonstrate approaches that would enable targeted communities to adapt to the potential impacts of climate change. It has four components: (i) integration of CCA into the agriculture and natural resources sectors, and strengthening the capabilities of relevant government agencies; (ii) assistance to poor rural communities to adapt to the effects of

climate change; (iii) improvement of access of end-users, especially in the agriculture and natural resources sectors, to more reliable scientific information for climate risk management; and (iv) support for project coordination functions through DENR's Foreign Assisted Projects Office.
- Climate Resilience and Green Growth in Critical Watersheds—this Asian Development Bank (ADB) technical assistance project funded by the Japan Fund for Poverty Reduction aims to assist CCC in demonstrating green growth and climate resilience alternatives in LGUs with critical watersheds. It covers nine municipalities in three watersheds: Lower Marikina, Camarines Sur, and Davao Oriental. It is expected to improve the ability of LGUs and other stakeholders to mainstream climate resilient and green growth options into development plans, programs, and policies.
- Millennium Development Goals—Achievement Fund (MDG-F) 1656: Joint Programme on Strengthening the Philippines' Institutional Capacity to Adapt to Climate Change—supported by the Government of Spain and UNDP. The Joint Programme is targeted to enhance capacities for CCA, mainstreaming climate risk reduction in developing plans and processes, and improving coping mechanisms. The program is also in support of the achievement of the MDGs.

While the above does not provide a comprehensive list of the adaptation programs and projects, it can be noticed that most projects are focused on building the adaptive capacity to perform CCA and DRR, particularly in mainstreaming and integrating these in development plans.

8.2.4 Local Climate Change Action Plan and Other Initiatives

Section 14 of the Climate Change Act stipulates the involvement of the local government units (LGUs) as the frontline agencies in the formulation, planning, and implementation of climate change action plans in their respective areas, consistent with the provisions of the Local Government Code, NFSCC, and the NCCAP. It fosters an inter-LGU collaboration, where barangays prioritize climate change issues and identify and implement best practices, with direct involvement of municipal and city governments. Municipal and city governments shall consider climate change adaptation as one of their regular functions, while provincial governments shall provide technical assistance, enforcement, and information management in support of municipal and city climate change action plans. LGUs are also directed to regularly update their respective action plans to reflect changing social, economic, and environmental issues. In the course of these actions, the national government shall provide technical and financial assistance to LGUs for the accomplishment of their Local Climate Change Action Plan (LCCAP).

In October 2014, the Department of Interior and Local Government issued Memorandum Circular 2014–135 outlining the guidelines on the formulation of the

LCCAP. It also aims to inform the process of mainstreaming climate change adaptation, including disaster risk reduction, in local mandated plans. The suggested process for the formulation of the LCCAP follows four stages: (i) preparatory stage (which includes organizing an LCCAP Core Team; (ii) data gathering, vulnerability assessment, risk analysis, and validation; (iii) planning, prioritization, and budgeting; and (iv) monitoring and evaluation.

In support of the above, numerous trainings were held to build the capacity of the LGUs in developing their own LCCAPS (http://calabarzon.dilg.gov.ph/index.php/news/279-lccap; http://region2.dilg.gov.ph/index.php/homepage/31-capdev-on-the-lccap-formulation-for-17-lgus-of-the-province-of-cagayan). At the time of writing, only the City of Iloilo has an available LCCAP online, which was completed in April 2014 and has a 14-year planning horizon (2014–2028). The plan follows a hazards-based approach (temperature increase, flooding, sea-level rise, typhoons, drought, and storm surge) in identifying adaptation and mitigation activities to climate change. Among the strategies identified were the development of green urban centers, infrastructures, structures, and open spaces; improving city flood management; mangrove reforestation; relocation of households in danger areas to safer resettlement sites; adjustment of the cropping calendar; and rainwater harvesting project; among others.

In line with the mandate set by RA 9729 to mainstream climate change in policy formulation, the Philippine departments have embarked on various initiatives to make their plans and programs climate change compliant and climate-proofed. For instance, Development Authority (DA) set up four strategic objectives that included the development of climate-smart agriculture and fisheries, and redefining of the Strategic Agriculture and Fisheries Development Zones (SAFDZ) considering climate change vulnerabilities (DA Memorandum S-01-13-0268). DOH has produced the Policy Handbook on Climate Change and Health that compiled policies on climate change and health to guide practitioners and other stakeholders in providing appropriate health services and infrastructures. The Department has also come up with the Philippines strategy on CCA for the health sector, which outlined actions related to governance, regulation, service delivery, and financing (DOH and WHO 2012).

DENR has implemented various programs on climate change, such as geohazard mapping, the National Greening Program, and adaptation to climate change in coastal areas. They have also incorporated CCA and DRR in the Philippines Environmental Impact System, through the issuance of Memorandum Circular 2011–005. In terms of technological and management service response, the Department of Science and Technology (DOST) has launched Project NOAH (Nationwide Operational Assessment of Hazards), heeding the call of Pres. Benigno Simeon Aquino III for a more accurate, integrated, and responsive disaster prevention. Part of the components of the project is setting up of monitoring stations and early warning systems across the country for disaster preparedness (http://www.gov.ph/programs/about-project-noah/).

The National Economic Development Authority (NEDA) mainstreamed CCA-DRR in the Philippine Development Plan 2011–2016 (Chap. 10: Conservation, Protection and Rehabilitation of the Environment and Natural Resources). Furthermore, the Housing and Land Use Regulatory Board (HLURB) developed Supplemental Guidelines on Mainstreaming Climate and Disaster Risks in the Comprehensive Land Use Plan. The guidelines have been used in the Project Climate Twin Phoenix.

Meanwhile, the MDG-F project has compiled best adaptation practices on climate change from 100 governments, business groups, and civil society groups (DENR 2012). While some of the adaptation strategies implemented are based on the CCA policy of the contributor, others were in the context of achieving their sustainable development goals. For instance, the business communities related how production or operation efficiency is improved and the contribution of their corporate social responsibility (CSR) projects not only in improving the livelihood of the communities but also in conserving the environment. Initiatives from the LGUs include the establishment of recycling facilities, banning the use of plastics and Styrofoam, and the creation of DRRM committees. Community-based actions, such as in disaster or emergency response, livelihood enterprises, rehabilitation, or protection efforts, are also among the common adaptation practices.

At the grassroots level, studies on assessing the adaptation of farming communities to climate change, particularly to extreme weather events and climate variability, revealed that the common adaptation strategies were diversification of livelihood (e.g., off-farm labor) and crops (e.g., drought-resistant crops) (Tapia et al. 2014; Lasco et al. 2010). In Pantabangan-Carranglan Watershed, other adaptation strategies recorded were reduced consumption, availing of loans, storing firewood, and food and asset disposal, among others (Lasco et al. 2010). Adaptation responses were observed to be mostly spontaneous and reactive in nature. They also merely address the immediate effects of a hazard and are largely characterized by bearing the losses from the impacts (Tapia et al. 2014). What is clear, however, is that the farming communities employ not just one but a bundle of adaptation strategies.

8.2.5 Research and Implementation Gaps

The Adaptation Knowledge Platform (2012) reviewed the provisions of the Climate Change Act vis-à-vis the Nairobi Work Programme (NWP), a mechanism under the UNFCCC to facilitate and catalyze the development and dissemination of information and knowledge that would inform and support adaptation policies and practices (https://www3.unfccc.int/). NWP defined nine work areas that would lead to the pathway of adaptation: (i) methods and tools; (ii) data and observation; (iii) climate modeling, scenarios, and downscaling; (iv) climate-related risks and extreme events; (v) socioeconomic information; (vi) adaptation planning and practices; (vii) research; (viii) technologies for adaptation; and (ix) economic diversification. The law was found wanting with regards to the role of climate science, climate scenarios, models, and economic diversification for adaptation as well as several aspects related to the socioeconomic dimensions of the Philippine society (Adaptation Knowledge Platform 2012).

The CCC, together with relevant stakeholders, also conducted a gap analysis of researches on climate change in the country to guide the development of a Research and Development Agenda on Climate Change (Climate Change Commission 2013). Following the NCCAP strategic priorities, water sufficiency and climate-smart industries lagged behind in research in terms of numbers. Most researches were focused on food security and ecological and environmental stability. Table 8.2 lists the identified gaps for the five NCCAP areas that specifically address CCA. Along with this, priority needs were also enumerated, which included enhanced capacity to establish and maintain observation facilities, zoning of settlement areas, identification of indicator species for health of habitat, and translation and popularization of research outputs.

The CCC has taken significant steps to close the identified gaps, particularly in building the capacities of the LGUs. In 2013, the Commission, with support from German Corporation for International Cooperation GmbH (GIZ), published the Compendium of Climate Change Vulnerability and Impact Assessment Tools to assist LGUs in integrating CCA and DRR in their local plans, particularly on how to conduct VAs (Balangue 2013). This is a product of the project "Supporting the CCC in the Implementation of the NCCAP" described previously. The implementation of eco-town also demonstrated how the local government could carry out the above process, with specific examples and outputs especially for the case of San Vicente, Palawan (CCC and GGGI 2014). DILG, in its MC 2014-135, has reinforced the process of mainstreaming CCA and DRR in development plans.

Of late, LGUs are working toward satisfying the requirement of completing their LCCAPs. The adaptation planning process involved in accomplishing this has already been laid down. How the adaptation strategies that were identified and prioritized would be implemented presents another story, as this is the area where the Philippines is usually constrained at different levels of governance. Therefore, there is a need to look into the actual practice of adaptation (Adaptation Knowledge Framework 2012).

The implementation of adaptation should also be true to its guiding principle of equity. Adaptation governance has to consider issues of equity in both the process and the outcome (Nelson et al. 2007). The former refers to "the fairness of the institutions, their representatives, and how they incorporate the diverse values and views of the community and the collective as well as the individual good", while the latter concerns with "the distribution of vulnerabilities across stakeholders within a population". In addition, aspects of ensuring successful stakeholder engagement that transforms practices at multiple levels are also critical for implementing CCA (Barletti et al. 2020). This will ensure that no one is left behind toward the vision of climate-risk resilient Philippines.

Table 8.2 Research gaps in the context of NCCAP strategies focusing on CCA

NCCAP strategic priorities	Identified gaps
Food security	• Lack of economic data to determine cost-effectiveness of practices • Vulnerability assessment (VA) studies mostly focused on rice and corn only • Lack of site-specific studies • Lack of harmonized VA tools • Lack of hatchery and breeding techniques for marine species • Lack of enabling mechanisms to upscale research results and translate scientific findings to policies
Water sufficiency	• Current research may suffice BUT need application of "climate lens" • Stand-alone research is needed according to technical competency of the institutions but coordinated research with other agencies is needed to achieve NCCAP targets • Few research have been conducted relating to adaptive capacity of water stakeholders
Human security	• Unavailability of high resolution maps for hazard and risk mapping • Lack of hazard to impact modeling capability • Inadequate data (existence, acquisition, accessibility, priority) • Inadequate funding • Non-existent peer review culture of completed research • Low degree of translating basic research to applied/operational mode • Fragmented research
Environment and ecological sustainability	• Lack of causal linkages with other strategic priorities and cross-sector contribution (e.g., location of corals, mangroves and their links to human security) • Weak causal attribution (e.g., looking into thresholds; specifying high, medium and low impacts in climate) • Mismatched risk and opportunities across sector • Unclear policy on the use of funds from overseas development assistance (ODA) as against the funds from government source for implementation of the action plans
Knowledge and capacity development	• No synergies with other policies (e.g., free, prior and informed consent on indigenous peoples or IPs) • Inadequate information on modes of climate financing • Low access to completed climate change research • Lack of policy/information on barriers and enablers to climate change data sharing and exchange • Low responsiveness of climate change research on needs of LGUs and vulnerable sectors (e.g., IPs) • Unclear use of climate change research (i.e., applied climate change research approaches) • Low capacity in translating scenarios into impacts • Few research works dealing with modes of engagement of stakeholders (e.g., complementation, collaboration, and cooperation) • Lack of modalities to popularize climate change • Low level of awareness and knowledge on climate change

8.2.6 Vulnerable Sectors and Regions

Key Concepts: Consistent with the IPCC Third and Fourth Assessment Reports, the Philippine Climate Change Act defines vulnerability as follows: "The degree to which a system is susceptible to, or unable to cope with, adverse effects of climate change, including climate variability and extremes. Vulnerability is a function of the character, magnitude, and rate of climate change and variation to which a system is exposed, its sensitivity, and its adaptive capacity". Vulnerability is therefore most often conceptualized as a function of exposure, sensitivity, and adaptive capacity (McCarthy et al. 2001). Exposure is the degree, duration, and extent to which the system is in contact with, or subject to, the perturbation (Gallopin 2006). Sensitivity is the degree to which a system is affected, either adversely or beneficially, by climate variability, climate change, or extreme events (IPCC 2007b). The effects may also be direct or indirect. Smit and Wandel (2006) distinguished exposure and sensitivity as almost inseparable properties of a system and are dependent on the interaction between the characteristics of the system and on the attribute of the climate stimulus. Meanwhile, adaptive capacity is the ability of a system to adjust to climate change (including climate variability and extremes) to moderate potential damages, to take advantage of opportunities, or to cope with the consequences (IPCC 2007b). The forces that influence the ability of the system to adapt, such as economic wealth, technology and infrastructure, information, knowledge and skills, equity, among others, determine its adaptive capacity (Smit and Wandel 2006; Ionescu et al. 2009). Exposure and sensitivity would result in potential impacts, and the impacts and adaptive capacity would determine the vulnerability (Fig. 8.5).

However, the emphasis changed with the release of the IPCC report in recent times (i.e., the AR5). The focus on vulnerability has shifted to climate risks, as the latter supports decision-making in the context of climate change and complements other elements (IPCC 2014). The change is also reflected in the new definition of vulnerability by IPCC AR5: the propensity or predisposition to be adversely affected (IPCC 2012). Vulnerability encompasses a variety of concepts and elements including sensitivity or susceptibility to harm and lack of capacity to cope and adapt. This new approach marries climate change adaptation and disaster risk management. With this new approach, vulnerability became a component of risk (from the disaster risk and management tradition), and it is the most palpable

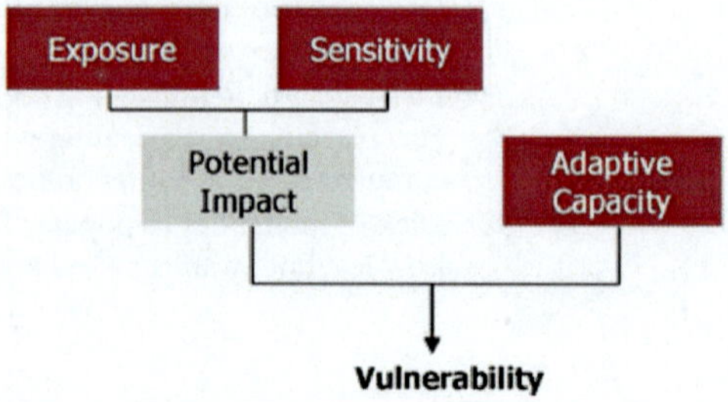

Fig. 8.5 Schematic presentation of vulnerability based on the IPCC's Third Assessment Report (TAR) and 4AR

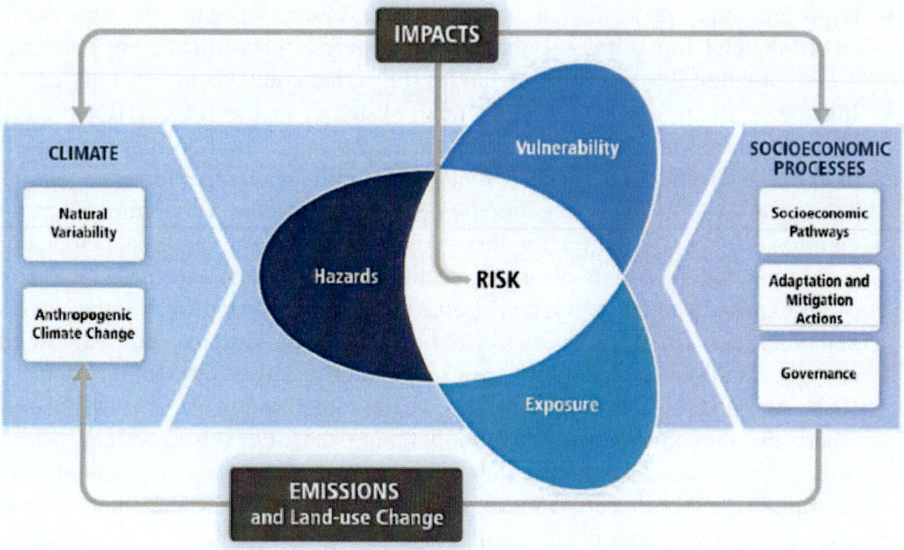

Fig. 8.6 Framework for the assessment of climate change impacts, adaptation, and vulnerability. *Source* IPCC (2014)

manifestation of the social construction of risk. Risk is interpreted using the below equation, as well as vulnerability (Mendoza 2014):

$$\mathbf{R = Hazard + Exposure + Vulnerability,}$$
$$\mathbf{where, Vulnerability = f(sensitivity + adaptive\ capacity)}$$

Figure 8.6 presents the illustration of the core concepts of SREX and AR5 reports. It evaluates the influence of natural climate variability and anthropogenic climate change on hazards that can contribute to risk. It also considers the role of the socioeconomic process in trends in exposure and vulnerability and its implications to risks (IPCC 2012). Thus, risk of climate-related impacts results from the interaction of climate-related hazards (including hazardous events and trends) with the vulnerability and exposure of human and natural systems. Changes in the climate system (left) and socioeconomic processes including adaptation and mitigation (right) are drivers of hazards, exposure, and vulnerability (IPCC 2014).

Aside from vulnerability (already defined above), the other important concepts that need to be elaborated are hazard, exposure, risk, and impacts. Their definitions are presented below (IPCC 2014):

- Hazard—the potential of a natural or human-induced physical event or trend or physical impact that may cause loss of life, injury, or other health impacts, as well as damage and loss to property, infrastructure, livelihoods, service provision, ecosystems, and environmental resources.

- Exposure—the presence of people, livelihoods, species or ecosystems, environmental functions, services, and resources, infrastructure, or economic, social, or cultural assets in places and settings that could be adversely affected.
- Impacts—effects on natural and human systems. In the IPCC AR5, the term *impacts* is used primarily to refer to the effects on natural and human systems of extreme weather and climate events and of climate change. Impacts generally refer to effects on lives, livelihoods, health, ecosystems, economies, societies, cultures, services, and infrastructure due to the interaction of climate changes or hazardous climate events occurring within a specific time and the vulnerability of an exposed society or system. Impacts are also referred to as *consequences* and *outcomes*. The impacts of climate on geophysical systems, including floods, droughts, and sea-level rise, are a subset of impacts called physical impacts.
- Risk—the potential for consequences where something of value is at stake and where the outcome is uncertain, recognizing the diversity of values. Risk is often represented as the probability of occurrence of hazardous events or trends multiplied by the impacts if these events or trends occur. Risk results from the interaction of vulnerability, exposure, and hazard.
- Adaptation—the process of adjustment to actual or expected climate and its effects. In human systems, adaptation seeks to moderate or avoid harm or exploit beneficial opportunities. In some natural systems, human intervention may facilitate adjustment to the expected climate and its effects.

It should be noted that since that shift in the focus of climate science from vulnerability to risk is just recent, most of the vulnerability assessment studies performed in the Philippines followed the old conceptualization of vulnerability being a function of exposure, sensitivity, and adaptive capacity, as somehow captured in the below section.

Vulnerable Sectors: While there have been quite a number of vulnerability assessments performed in the Philippines, the program called "Adaptation to Climate Change and Conservation of the Biodiversity in the Philippines" (ACCBio) has already come up with an assessment of the vulnerability of different sectors in the Philippines, which also formed the basis for adaptation actions in the country, as well as the identification of gaps and needs for this scoping study. Much of the assessment highlighted the potential impacts of future climate on different sectors in the Philippines (Table 8.3). The gaps and issues as well as required actions were also highlighted.

While the reports of the ACCBio and even the priorities of the NCCAP have focused on sectoral vulnerabilities, it should be noted that all the above sectors are interconnected and should be dealt with holistically, to achieve the NFSCC vision of a climate risk-resilient Philippines. Recent national projects in the country, such as Monitoring and Detection of Ecosystems Changes for Enhancing Resilience and Adaptation in the Philippines (MODECERA), are making a serious attempt to adopt the landscape approach. This approach lends a framework for applying systems thinking in addressing problems related to climate change. It highlights the biophysical, social, economic, institutional, and political components of the

Table 8.3 Vulnerability to climate change of different sectors in the Philippines. *Source* ACCBio (2009)

Sector	Sector profile/vulnerability	Gaps and issues
Agriculture	• Agriculture contributes 18% to the country's GDP and employs about 30% of the country's labor force. One-fifth of the country's economy is rooted in the agricultural sector which contributes 18% to the country's GDP. Thirty percent of the country's labor force is in this sector • Almost 80% of the best arable land, worked by poor farmers, produce rice, corn, and coconut • Soil erosion and land degradation affect almost half the land area used for cultivation, leading to higher input requirements • High temperatures, wet weather conditions often affect biological and production cycles of plants and fruit trees, thus changing fruiting and harvesting dates • Problems relating to pests and diseases in rice, scale insects in fruit trees, and invasive weeds have emerged; this is the result of movement caused by various flood events in the uplands	• National research agenda and information from research limited for various crops and livestock • Conversion of prime agricultural land

(continued)

Table 8.3 (continued)

Sector	Sector profile/vulnerability	Gaps and issues
Biodiversity	• The Philippines is one of the 18 "mega" biodiversity countries in the world. It is ranked fifth in terms of diversity in plant species and fourth in native bird species • About 25 genera of plants and 49% of wildlife are endemic, while inland waters host 121 endemic and 76 threatened freshwater species • Human-induced and direct stress on species and ecosystem diversity and climate change threatens biodiversity • The clearing of forests, expansion of agricultural lands, development of settlements, urban development, water and air pollution, mining and destructive fishing methods, and economic activities which are not environment-friendly threaten resources • Many of the foregoing critical threats will further be exacerbated by climate change, which would have adverse impacts on the integrity of species and ecosystem diversity • Wildfires had been more frequent during drought conditions, destroying large areas of forest • The spread of invasive alien species of plants, insects, and pests in natural ecosystems has been attributed to the extreme swings in temperature and moisture availability in these areas • There is less than 6% of intact coral reef left due to degradation resulting from destructive fishing practices and heavy siltation • As coastal water temperatures increase, coral reefs will become bleached • Seagrass meadows are slowly drying out when exposed during low tides	• The Philippine Convention on Biological Diversity (CBD) Report of 2009 noted the absence of national baselines, standards, and indicators for monitoring progress in implementing biodiversity conservation programs • Manpower and financial constraints result in poor management of protected areas • Notable lack of action plan to address the problem of alien invasive species. Expansion of plantation forests in forestlands displacing native species of forest and dearth of local scientific researches measuring the impact of climate change on biodiversity. As plantation forests take over natural forest lands, native species are reduced; therefore, an action plan must address this problem of alien invasive species • There is a lack of research and monitoring plans on climate change impacts on species and ecosystem diversity • Weak institutional capacities to plan and implement program on biodiversity adaptation to climate change must be addressed • Integration of biodiversity conservation development is needed, particularly in the development of human settlements and towns and cities which have simplified much of the landscape and reduced or obliterated ecosystem diversity

(continued)

Table 8.3 (continued)

Sector	Sector profile/vulnerability	Gaps and issues
Coastal and marine	• Coastal and marine ecosystems of the country consist of beaches, mangrove forests, coral reefs, seagrass beds, soft-bottom communities, open marine waters, and small islands • Beach forests and mangrove forests have been largely cut down to give way to human settlements and other purposes. It is estimated that less than 100,000 ha of original mangrove areas, estimated in 1918 at half a million ha, remain today • Marine turtle nesting sites on beaches have been reduced in area, and mining of beaches for minerals threatens to destroy considerable areas. These disturbances have reduced the capacity of these ecosystems to buffer strong wave action and storm surges • Pollution has affected all coastal ecosystems: the absence of good forest cover compromised the land resulting in erosion and sediment flow from the higher terrain and also dredging and land reclamation in coastal areas • The coastal waters and coral reefs have been affected by rising sea surface temperatures, resulting in episodes of coral bleaching that have had some impact on the fisheries of coral reefs • Rising sea water temperatures and increasing acidity threaten larval survival and transport in open oceans. The normal oceanic productivity via the plankton will be adversely affected, as projected. The decrease in basic productivity of open marine waters will be felt throughout the entire food chain leading to humans • As the frequency of storm surges increases, the above situation will worsen as a result of the decrease in the population of fish and other organisms • Coastal communities are already experiencing the effects of sea-level rise. Flooding brought about by heavy monsoon, strong winds and storm surge accompanied by high tide, cause flooding and displace communities and destroy infrastructure	• The establishment of fully protected marine areas is urgently needed up to the extent that about 30% of the coastal and marine ecosystems shall have acquired full protection and management

(continued)

Table 8.3 (continued)

Sector	Sector profile/vulnerability	Gaps and issues
Energy	• The main concern of the Philippines energy sector about climate change relates to its mandate of ensuring an adequate supply of oil, power and gas for domestic, transport and industrial use to drive the country's economy • The country currently relies on fossil fuels to generate power and to run machines for transport and manufacturing • Hydropower generation is sensitive to the amount, timing, and geographical pattern of precipitation as well as temperature. Reduced stream flows are expected to jeopardize hydropower production • Changes in water policies can be influenced by changes in precipitation and runoff patterns. With heavy rain and heavy stream flows, water has to be diverted from hydropower facilities. This is to prevent damage to the dams, or water may be released from reservoirs to avoid flooding. Photovoltaic (PV) electricity generation is suitable in the country, but solar radiation (i.e., the energy source for these systems) may be affected by climate change, resulting in increased cloudiness and decreased levels of daily global radiation availability • Wind power generation is susceptible to variations in ambient temperatures, humidity, and precipitation. Increased variability in wind patterns could create additional challenges for accurate wind forecasting for generation and dispatch planning, for the siting of new wind farms, and for the integration of wind with the utility grid • Changing climatic conditions alter historical patterns of precipitation and runoff which may complicate operations of existing thermoelectric power plants • Power plants located along the coast are subjected to the possible effects of sea-level rise and coastal storm surges • Power transmission and distribution systems may be affected by several aspects of climate change: sea-level rise, increased temperatures, and extreme typhoon events	• The energy sector faces two main challenges as a result of the changing climate: (1) responding to the increase in energy demand as a result of increasing temperatures; and (2) the ability of energy infrastructure (e.g., power plants, refineries, depots, transmission, and distribution system) to adapt to changing climatic conditions • Need to assess the energy systems/infrastructure vulnerability, pressure in energy demand, and energy supply vulnerability • Need to develop models on climate change impacts including weather extremes, variability, hydro resources, mean temperatures, and wind speeds to allow energy experts to assess the implication of climate on demand and supply of energy • Vulnerability assessments of the energy systems, e.g., power generation, transmission and distribution, fuel production, and transport, which are currently lacking will help identify risks in relation to extremes and variability

(continued)

Table 8.3 (continued)

Sector	Sector profile/vulnerability	Gaps and issues
Forestry	• Forest ecosystems are vital to human survival and stability of cultures; they support and regulate natural processes like the maintenance and regeneration of biodiversity, soil formation and nutrient cycling, crop pollination, infiltration of water, and enhancement of microclimate • Current forest cover is 7.1 million hectares. Global warming caused by high amounts of GHGs in the atmosphere is a result of deforestation and forest degradation activities like logging, slash-and-burn farming, forest fires, and the conversion of forestlands to agricultural lands, human settlements, and other development efforts • Climate change exacerbates deforestation and forest degradation. Some evidences include forests dry up and become more vulnerable to destructive fires; reforestation efforts become less successful; species migrate or perish; wetlands alternately dry up or flood up; and mangroves are driven inland	• Information and research gaps on the extent of environmental degradation and related bio-physical impacts, and the forest ecosystem's response to a changing climate while interacting with other environmental modifiers • Potential of forestry and agro-forestry systems in mitigation and adaptation, e.g., carbon sequestration rates of Philippine trees especially in various agro-ecological zones of the country • Continuing elusiveness of effective forest stewardship and sustainable forest management systems. Only a few instances of success in forest stewardship can be seen in spite of several million hectares of forest land awarded
Health	• Direct effects of climate change on health come from variable climate events such as floods and heat waves which may cause death • Indirect effects come through disease transmitted by secondary vector-borne hosts • Health impacts of climate change mediate effects on other sectors as much as it is mediated in turn by them • The disease occurrence triad of environment-host-reservoir should be considered when discussing the impacts of climate change on human health as the relationship of these three shows the dynamics of disease transmission • Eight of the ten leading causes of morbidity in the Philippines are infectious in nature (pneumonia, diarrhea, bronchitis, influenza, tuberculosis, malaria, chicken pox, and measles • Mosquito-borne diseases, such as malaria, dengue, and filariasis, are present dangers in endemic areas. The surge in dengue cases occurring in cyclical outbreaks every three to five years remains a threat to public health	• Need to look at the possibility of climate-sensitive diseases being caused, if not being exacerbated, by changes in the climate which may alter their life cycles and make their existence more favorable and extended, become more virulent, travel more distances, and therefore make transmission to humans more likely as to cause infirmity

(continued)

Table 8.3 (continued)

Sector	Sector profile/vulnerability	Gaps and issues
Infrastructure	• Prolonged and frequent rainfall, strong winds, temperature variations (extreme heat and cold) among others can lead to accelerated structural fatigue and materials failure, as well as greater demands on flood control and drainage, construction, and operation and maintenance needs • Sea-level rise and storm surges are likely to impact operating and future infrastructure systems. Published reports indicate a one-meter sea-level rise could make 703 municipalities vulnerable • Variabilities in rainfall and rainfall extremes can have serious consequences, especially with typhoons causing flooding, landslides, accelerated erosion, and sedimentation. Over time, this will cause the collapse of infrastructure or less efficient functionality accompanied with higher maintenance or rehabilitation costs • Mean temperatures have been increasing in the country and could result in accelerated structural fatigue and materials failure in vital infrastructures such as roads and bridges	• The infrastructure sector is heavily dependent on climate parameters for planning, design, and implementation. Currently, the access to reliable baseline hydrological and meteorological information is hampered by the absence of longer record-keeping due to a variety of reasons, including budgetary constraints and obsolete information and data gathering equipment • The engineering and building codes and standards adopted in the country were developed when climate change was not yet widely acknowledged nor confirmed. Codes and standards for infrastructure have to be revisited to take into account risks that are likely to damage with climate over an extended timeline

(continued)

Table 8.3 (continued)

Sector	Sector profile/vulnerability	Gaps and issues
Water	• Rainfall, surface water (rivers, lakes, and reservoirs), and groundwater provide the Philippines with abundant freshwater • Annually, the country gets an average of 965–4,064 mm of rainfall. There are 12 water resource regions and 421 river basins, of which 20 are considered major comprising about 990 km^2 of the basin area. The dependable surface water supply from rivers, lakes, and reservoirs is estimated at 125.8 billion cubic meters. Groundwater potential, on the other hand, is approximately 20.2 billion cubic meters • Agriculture and households are the largest consumers of water, accounting for 52.22% and 32.93%, respectively • The changes in water quantity and quality due to climate change are expected to greatly affect food production, availability, stability, access, and utilization, which will consequently lead to decreased food security and increased vulnerability of the poor • Extreme climate variability will affect the function and operation of the country's existing water infrastructure—including hydropower, structural flood defenses, drainage, and irrigation systems • More efficient water management practices will be needed to cope with water supply reliability, flooding risk, health, agriculture, energy, and the aquatic ecosystems • An increase in population, urbanization, industrialization, and changing irrigation demands will greatly stress the existing water management system and structure	• The national and local capacities in climate change adaptation and integrated water resources management remain inadequate for supporting a management framework and potential adaptation strategy • The gap in research on water resources supply and demand is also reflected in the lack of updated scientific information in the sector, leading to conflicts in the implementation of water resources policy and poor regulation of resource use and development • Inadequate awareness on water consumption efficiency and technological innovations lead to wasteful use of already scarce resources • Gaps in sector financing and investment also hamper effective and efficient delivery of water services

system. Central to addressing the vulnerability of each sector is enhancing the adaptive capacity of communities, institutions, and other stakeholders. Local studies have highlighted high dependence on primary industries (agriculture, fishery, and forestry), inadequate access to basic services (e.g., education, health, and finance mechanisms), and low level of information and awareness as sources of vulnerabilities (Tapia et al. 2014; Toda et al. 2015). This necessitates enhancing education and building the capacity of the stakeholders, specifically addressing the different social weaknesses for each sector as indicated in Table 8.3.

Vulnerable Regions: The different risk factors in terms of climate hazards in the Philippines are presented in Fig. 8.7 based on (DENR 2012). It can be noted from the below map that there are five major climate-related hazards that the Philippines is at risk to sea-level rise, extreme rainfall events, increased heating events, increased ocean temperatures, and a disturbed water budget. It can also be deduced from Fig. 8.7 the different system components or services that would be affected by such hazards, with the highly populated and economic centers, Metro Manila and Region IV, exposed to Cluster III hazards (i.e., extreme heating events, disturbed water budget, and sea-level rise); the agricultural area in Region III exposed to Cluster IX hazards (i.e., extreme heating events, extreme rainfall events, disturbed

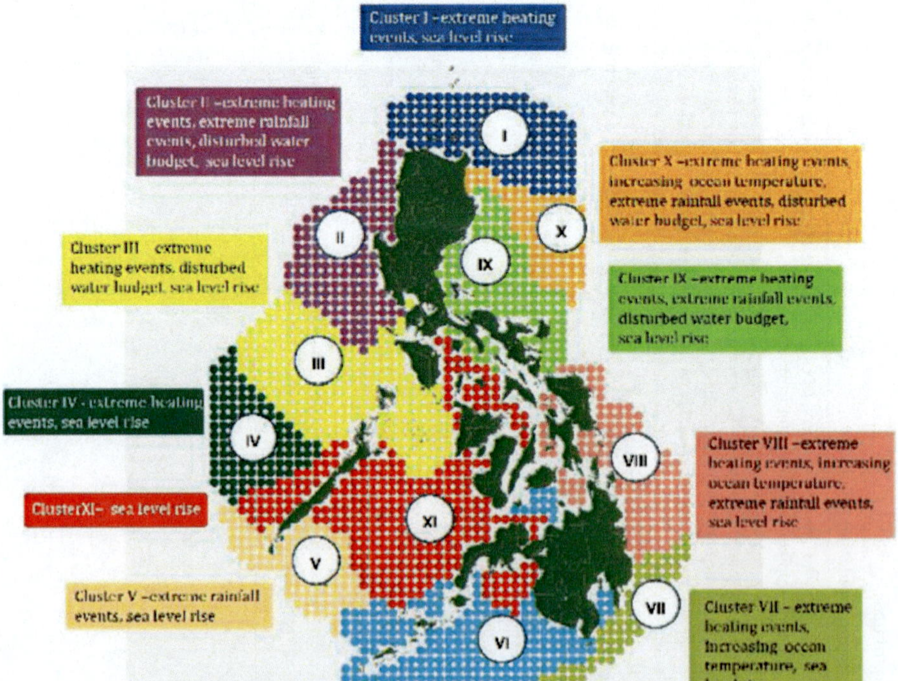

Fig. 8.7 The philippines exposure map on climate change. *Source* DENR (2012)

water budget, and sea-level rise); and the island provinces in Southern Luzon and Visaya exposed to Cluster XI hazard (i.e., sea-level rise).

8.3 Adaptation Priorities

8.3.1 Approach

The Climate Change Commission (2013) has conducted a series of consultations on developing a research and development agenda for climate change in the Philippines, and this has resulted in the publication of the "Philippine Research and Development Agenda in Support of the National Climate Change Action Plan Strategies Priorities". This report, consistent with the identified priorities in the National Climate Change Action Plan (NCCAP), included a gap analysis following an inventory of climate change research that has already been done in the country, areas that need further studies, as well as new and emerging issues. This Research and Development Agenda is also intended as the guiding document for the country's research, especially for the science and academic institutions assisting the public, private, and non-government sectors in the quest for formulating solutions to issues and concerns brought about by climate change.

Figure 8.8 presents one of the outputs of the climate change research inventory, indicating that among the seven thematic priorities identified in the NCCAP, environmental sustainability and food security were the areas with the most number of research conducted. Other priorities such as water sufficiency, human security, sustainable energy, and climate-smart industries and services need to catch up. Nevertheless, it is not only the number that matters but also the specific gaps and priority needs that need to be addressed under each thematic area. Table 8.4 lists the gaps and priority needs identified in the Research and Development Agenda for climate change, under the five NCCAP priorities that specifically tackle climate change adaptation.

The gaps and priorities for climate change research and development as well as the vulnerabilities of each sector and the number of current research already conducted for the different thematic areas served as the basis for the identification of several thematic areas for local action-oriented pilots (Table 8.5). The findings were then shared with stakeholders to identify potential pilots.

A local action pilot project was not proposed for the theme under food security because there are already ongoing projects that address this (such as the Smarter Approaches to Reinvigorate Agriculture as industry in the Philippines or SARAI Project). This is to avoid duplication of efforts and to redirect resources to other areas with critical needs. In its place, a cross-cutting theme on governance was identified as another priority, especially in support of the current initiative in the country for the local government units (LGUs) to formulate their Local Climate Change Action Plan (LCCAP).

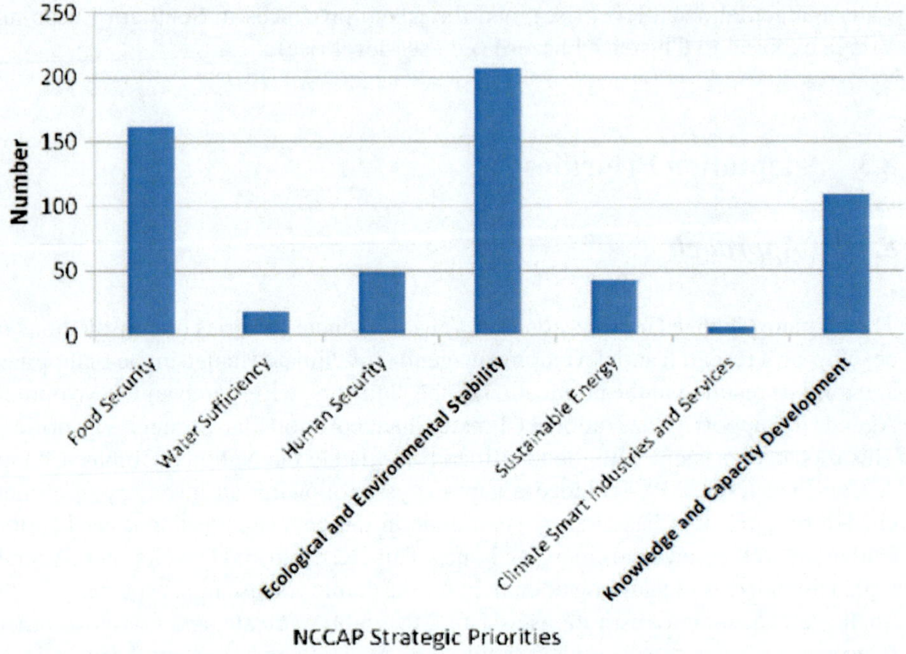

Fig. 8.8 Number of research conducted in the country under the strategic priorities of the NCCAP. *Source* CCC (2013)

While the participants in the consultation-workshop agreed that the proposed local action-oriented research pilots were needed to address current needs in the country, it was, however, observed that the priority areas are linked together, and realizing the objective of one area would mean addressing the pressing concerns of another. Hence, it was decided to select three areas representing the islands of Luzon, Visayas, and Mindanao, where the project would be implemented. It was also highlighted to select an area where adaptation in an island ecosystem, which is the case for many LGUs in the Philippines, would be demonstrated. Each of the local-oriented adaptation projects would have different components to holistically address the needs or vulnerabilities in the chosen site. It was also thought useful that some sites already have a Local Climate Change Action Plan to demonstrate its implementation. Based on the above conclusion, the following potential projects were identified in order of priority.

- Enhancing community resilience in San Vicente, Palawan through effective implementation of LCCAP with a focus on human security and capacity development. This involves strengthening the capacity of the LGU to implement LCCAP, improving human security through the development of appropriate livelihoods and environmental programs, and developing capacity and ensuring sustainability through the establishment of community-based monitoring system.

Table 8.4 Research gaps and priority needs in the context of the NCCAP

NCCAP strategic priorities	Identified gaps	Priority needs
Food security	• Lack of data on the cost-effectiveness of mitigation and adaptation practices • Vulnerability and Adaptation (V&A) studies mostly focused on rice and corn only • Lack of site-specific studies • Lack of harmonizing VA tools with high precision • Lack of hatchery and breeding techniques for marine species • Lack of interface with industry/community-based enterprises to upscale research results • Lack of enabling mechanisms to translate scientific findings to policies	• Enhance capacity to establish and maintain observation facilities to collect data to be used for climate change protection and modeling • Studies on cost and benefit of adaptation and mitigation technologies • Development of site-specific climate-resilient crops, livestock, and aquatic species • VA focus on other commodities (high-value vegetables, other staples) • Enhanced capacity to establish and maintain observation facilities to collect data to be used for climate change projection and modeling • Awareness campaign and capacity building on climate change mitigation and adaptation practices • Studies on climate-resilient crops (other than rice), livestock, and aquatic species • Studies on emerging pests and diseases under changing climate • Studies on the analysis of food production areas versus biofuel production areas in the Philippines vis-à-vis climate change impact • Enabling mechanisms to translate scientific findings to policies • Development of linking strategies with the industry/community-based enterprises to upscale climate change-related technologies • Development of harmonized VA tools with high precision

(continued)

Table 8.4 (continued)

NCCAP strategic priorities	Identified gaps	Priority needs
Water sufficiency	• Current research may suffice BUT need application of "climate lens" • Stand-alone research is needed according to the technical competency of the institutions but coordinated research with other agencies is needed to achieve NCCAP targets • Few research works have been conducted relating to the adaptive capacity of water stakeholders	Vulnerability, impacts, and adaptation (VIA) of water sources • Groundwater constrained areas (e.g., Cebu City and other island provinces) • Major river basins • Water infrastructure for major cities • Multi-purpose dams
Human security	• Availability of high-resolution maps • Impact modeling per sector • Data (existence, acquisition, accessibility, priority) • Funding • Peer review culture • From being basic research to applied/operational • Fragmented research	1. Focus more on policy work • Climate change and women, children, health • CCA-DR risk transfer mechanism • Zoning/settlement areas • Responsiveness of social security program to adverse effects of climate change 2. Technical research • Baseline data • Early warning system • Biogeography • Urban design • Evaluation of mosquito diversity in sentinel sites in the Philippines in relation to disease prevention and control • Coordinated research with NDRRMC on DRR and CCA linkages

(continued)

Table 8.4 (continued)

NCCAP strategic priorities	Identified gaps	Priority needs
Environmental and ecological stability	• Lack of causal linkages with other strategic priorities and cross-sectoral contribution (e.g., location of corals, mangroves, and their links to human security) • To look more into thresholds (e.g., specifying high, medium, and low impacts in climate) • Matching of risk and opportunities across sectors • In terms of policy, reconciliation of ODA with the action plans of the government. There is a need to map out areas of interest and donors, the gaps to be filled up by the government (funds, area, and research topics)	1. Climate change mitigation and adaptation strategies for key ecosystems developed and implemented • VIA on selected ecosystems in Central Luzon • Climate change vulnerability mapping and geospatial analysis in Cagayan River Basin • Comprehensive VA of key coconut production areas • Climate sequestration of potential areas for coconut expansion • Crop suitability analysis based on climate variables • Region-wide VA of near-shore, marine, coastal, and small island ecosystems • Incorporating climate risks in land use plans of LGUs • Computer-based interactive-simulated climate change scenario for communities and schools • Assessment of the impact of NGP on poverty alleviation and greenhouse gas mitigation 2. Management and conservation of protected areas (PAs) and key biodiversity areas (KBAs) • Biodiversity species vulnerability assessment (impacts on endangered species) • Biodiversity mapping and management • Wetland conservation • Establishment of permanent biodiversity plots to monitor habitat changes • Threshold level studies in KBAs (endangered species and commercially important species) • Identification of indicator species for health of habitat • Study of the effects of slash and burn farming on biodiversity • Management of river systems • Rainforest farming and riparian zones, green corridors for biodiversity conservation in the second district of Nueva Ecija 3. Environmental laws strictly implemented • Assessment of logging and mining permits issued by DENR (2009–2012) • Inventory and assessment of municipal environmental codes • Assessment of fisheries code implementation • Interface of customary laws with written laws • Solid waste management • Strengthen the establishment of material recovery facilities (MRFs) in each barangay in Metro Manila • Unified vehicular volume reduction program • Bus tagging/e-tagging 4. Capacity for integrated ecosystem-based management approach to PAs and KBAs • Mainstreaming of climate change in municipal planning for timing and usability • Community-based rural eco-agri-tourism planning and programming • Capability building in response-ability and early warning systems of municipalities • Training on fund sourcing for implementation of environmental and PA laws • Training on fund sourcing for implementation of fishery laws • Environmental awareness and education on climate change impacts and disaster risk reduction 5. Natural resource accounting institutionalized • Wealth assessment and valuation of ecosystems services project • Studies onthe use of benefits transfer method in setting up quotas for collection of natural resources • Analysis of the time series data available on natural resources and determination of its significance to affected communities • REDD + baseline

(continued)

Table 8.4 (continued)

NCCAP strategic priorities	Identified gaps	Priority needs
Climate smart industries and services	• Vulnerability analysis • Information, education, and communication • Capacity building • Appropriate technology development and/or transfer	• Development and/or transfer of appropriate technology • Development of financing and incentive packages • Support packages for transition (e.g., safety nets, innovation support) • Development of eco-efficient standards for products and services

Source CCC (2013)

Table 8.5 Proposed areas for action-oriented pilots

Priority area	Proposed action-oriented pilots	General objective
Water sufficiency/water security	Risk assessment of the impacts of extreme events to multi-purpose dams (e.g., San Roque) and host/downstream communities	To assess the risk associated with extreme events on dam infrastructures, operations, and response strategies (e.g., release of water) and their impacts on host/downstream communities
Human security	Local level or community-based monitoring of impacts and adaptation	To establish a community-based monitoring system to support adaptive management and address lack of data
Knowledge and capacity development	Innovative climate-smart community information and learning centre	To pilot an innovative climate-smart information and learning centre managed by a community
Ecological and environmental sustainability	Forestry: Climate change adaptation and mitigation strategies through community forestry	To explore how community forestry can further contribute to CC adaptation and mitigation while taking advantage of mechanisms such as REDD+
Cross-cutting: governance	Action research on strengthening capacity of LGUs to implement LCCAP	To enhance capacity of LGUs in the implementation of the LCCAP (proposed area: San Vicente, Palawan)

- Reducing Vulnerability to Climate Change toward Resilient Ecosystems and Communities in the Upper East Coast Watershed of Davao Oriental (Baganga, Boston and Cateel). This involves addressing water security through effective watershed management, management of production and protected forests through community forestry, taking advantage of mechanisms such as REDD+, and protection of coastal and marine resources.
- Enhancing Climate Change Adaptation in a Watershed Within an Island Ecosystem: The Case of Catubig Watershed. This involves improvement of water security through enhanced water governance, building the capacity of the communities through improved livelihood strategies and environmental management, and the establishment of an innovative climate-smart community and information center.

The proposed pilot projects support national priorities and build on existing initiatives toward CCA in the country, particularly that of the formulation of the LCCAP by the LGUs and the Eco-town Project of the CCC which aims to integrate and mainstream CCA and disaster risk reduction (DRR) in local development plans, programs, and activities to promote green growth. Furthermore, a landscape approach will be followed in designing holistic adaptation responses that focus not only on the strategies but also on the process to address the needs and priorities in the proposed study site. Landscape approach is generally described as "addressing social-ecological systems at the landscape scale; related to resource management and/or environmental goals; and framed around the concept of multifunctionality, with the aim of achieving multiple objectives through the approach" (Freeman et al. 2015). This approach is also characterized by the concepts of multifunctionality, transdisciplinarity, participation, complexity, and sustainability.

In designing adaptation strategies to build resilience, the Climate Resilience Framework (CRF) would be applied which focused on building the adaptive capacity of the sub-components of the system, namely the agents, institutions, and systems (infrastructure and ecosystems). Agent refers to people and their organizations, whether as individuals, households, communities, public and private sector organizations, or companies. Characteristics that need to be enhanced for the agents are responsiveness, resourcefulness, and capacity to learn. Institution refers to the rules, norms, beliefs, or conventions that shape or guide human relations and interactions, access to and control over resources, goods or services, assets, information, and influence. Capacities to be developed in an institution are access rights and entitlements, decision-making processes, information flows, and applications of new knowledge. Lastly, systems are considered as the combination of ecosystems and infrastructures. Characteristics of ecosystems and infrastructure systems that contribute to resilience include flexibility and diversity, redundancy and modularity, and safe failure (ISET 2013).

8.3.2 Proposed Local Level Pilot

The municipality of San Vicente, Palawan is a first-class municipality located on the northwestern side of the province and 186 km from Puerto Princesa City (Fig. 8.9). It occupies a total land area of 165,797 ha and has jurisdiction over 10 barangays, namely Alimanguan, Binga, Caruray, Kemdent, New Agutaya, New Canipo, Poblacion, Port Barton, San Isidro, and Sto. Niño. It has a total population of 30,565 based on the 2010 Census, with a growth rate of 6%. Based on the Community-Based Monitoring System (CBMS) survey, the municipality's population increased by 3,500 from 27,065 in 2008. The total number of households is 6,460 with an average size of five members. San Vicente has 22 islands and islets within its municipal boundaries serving as tourist attractions. Its largely untapped beaches, coral reefs, waterfalls, forest cover, and mangrove areas are home to 23 of the 25 wildlife species found in Palawan. It has a forest area of 83,080 ha. San Vicente also hosts 24 ethno-linguistic resident groups, each with their own distinct dialects and cultural heritage. The main livelihoods in San Vicente are fishing (46.98%) and farming (37.43%). Rice and coconut are the major agricultural crops. As a first-class municipality, San Vicente aspires to tap

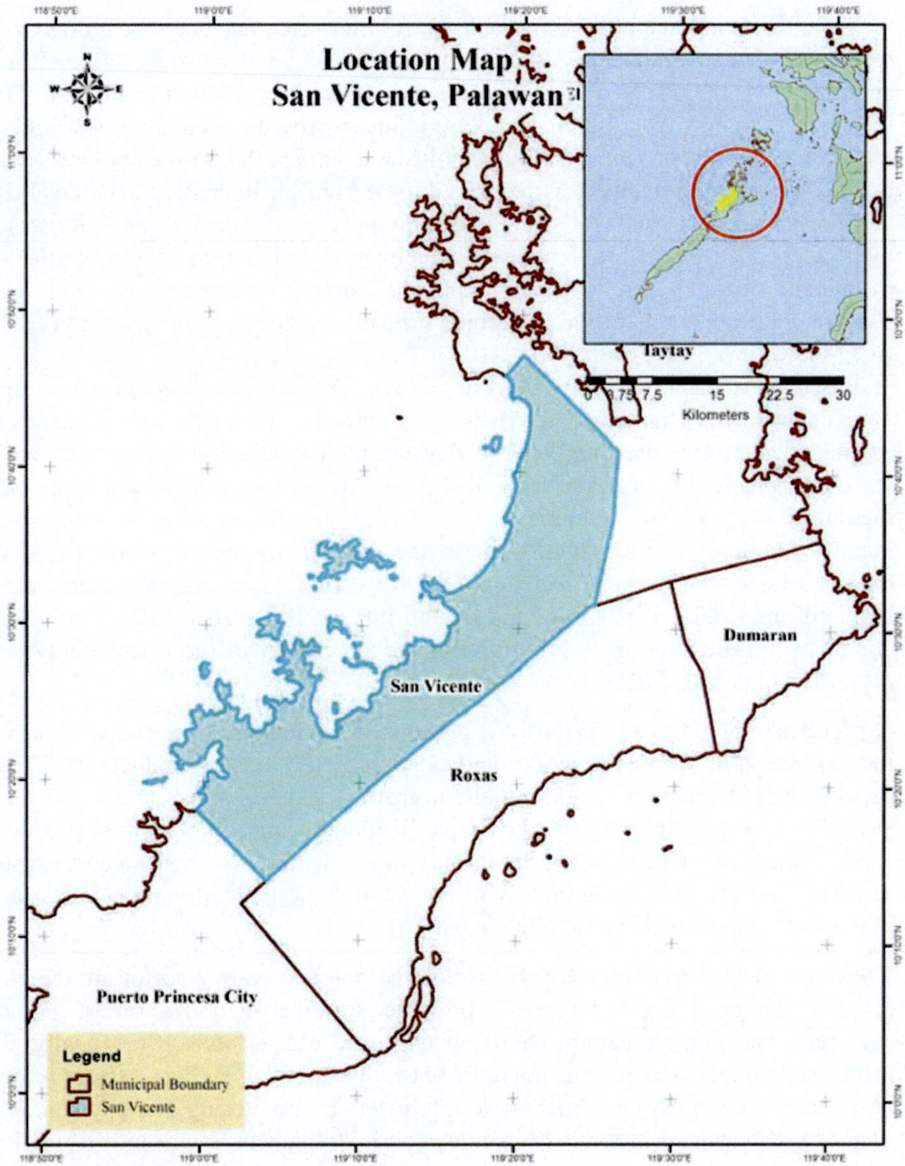

Fig. 8.9 Location map of San Vicente, Palawan

the potential tourism, particularly the 14-km long beach from Poblacion to Alimanguan, waterfalls, dense forest cover, and rich fishing ground to generate more income and attract investments. Given its rich natural resource endowments, the municipality aspires to be a top tourist destination that strikes the balance between economic growth and environmental sustainability.

Since the landscape approach would be followed in enhancing the climate risk resilience of the municipality, this proposed project shall consider not only the administrative boundary of San Vicente but also its ecosystem boundary. This means that systems that influence and are influenced by the municipality would be included in the assessment. The municipality was among the prioritized sites (i) as, being a coastal community, it is perceived to be highly vulnerable to sea-level rise and coastal flooding; and (ii) people's livelihood (farming, fishing, and tourism) is highly dependent on natural resources, which are already being adversely affected by climate change, thereby exacerbating the poverty situation in the area. San Vicente, Palawan is one of the Philippine project sites for the demonstration of the CCC's Eco-town framework.

Based on a comprehensive analysis of San Vicente's development priorities, natural resources, hazards and risks, and climatic vulnerability, a menu of adaptation measures was put forward through the consultation between the local stakeholders and the local and foreign experts. The task of prioritizing adaptation measures is critical to planning due to limitations in resources to finance all identified options. Through multi-criteria analysis, several options were prioritized based on effectiveness, cost, technical feasibility, social and cultural acceptability, required time, and sustainability and overall impact. It is also worth noting in the course of prioritization that the current adaptive capacity of the municipality was also taken into consideration.

Agriculture: Some of the prioritized options for agriculture are grounded on the modernization of farming practices and technology to increase productivity. These include the construction of additional automatic weather stations and small-scale irrigation facilities; training on alternative livelihood (non-timber forest product); establishment of farmers' field schools and programs; introduction of new climate-resilient crop varieties, including hybrids; and alteration of cropping calendar and practices to adjust to changing climate patterns.

Coastal and Marine: Top measures for the coastal and marine sector are targeted to save the depleting coral reefs, promote sustainable utilization of marine resources, and prepare natural catastrophes in the coastal areas. Specifically, the proposal highlights the establishment of sea walls and dikes in Port Barton; setting up of early warning system; mangrove reforestation; monitoring of fish catch, also illegal fish catch practices; organizing and strengthening fisherfolk bodies; undertaking herbivore seeding and creating the necessary mix of marine habitat types to enhance coral resiliency; alternative livelihood training; drawing in private sector in coastal planning and management; conducting training and orientation programs on disaster risk reduction and management, and the like.

Health: The adaptation measure for health seeks to address the primary vulnerabilities specifically the climate-induced diseases and hazards. The prioritized options: the provision of clean and adequate water supply system, the use of insecticide on bed nets, and rapid treatment strips for malaria and water tablets. As preventive measures: promote and monitor regular health checks;

provide training on early detection/treatment of infectious, water/vector-borne diseases; public health and hygiene training for the youth; and increased enrolment in health financing facility (PhilHealth) are paramount.

Based on the vulnerability and adaptation assessment performed through the Eco-town Demonstration Project, farming and fishing were the major livelihood sources and also formed part of the municipality's vulnerabilities. Nevertheless, San Vicente also has a high potential for community-based ecotourism, of which the forestry and marine sectors are the natural assets. This is also hoped to safeguard the forestry and marine sectors from threats of degradation. However, it can also be noticed that the vulnerability and adaptation assessment performed is lacking on the social component, particularly that of stakeholder analysis, institutional analysis, organizational capacity analysis, and the like. Results from such endeavors would be particularly useful in the facilitation of adaptation strategies.

Taking the municipality of San Vicente as one socioecological system, this proposed project adopts a holistic approach at adaptation toward building resilience to climate change, keeping in mind the concepts of multifunctionality, transdisciplinarity, participation, complexity, and sustainability and specifically targeting the sub-components of resilience. The project could be implemented by the University of the Philippines Los Baños (UPLB) and the Department of Environment and Natural Resources (DENR), with partnership from the Municipal Government of San Vicente, Palawan. Since San Vicente was a recipient of the Demonstration of the Eco-town Framework by the Climate Change Commission, collaboration from the latter would be sought to build on their previous and current initiatives. Relevant government agencies and other institutions (such as SEARCA, ASEAN Centre for Biodiversity, etc.) would also be tapped in fleshing the details of the project activities and their implementation. Additional funding could also be requested from the Philippine Council for Agriculture, Aquatic and Natural Resources Research Development (PCAARRD). The proposed project is expected not only to contribute towards improving the resilience of San Vicente, Palawan, but also in potentially developing a framework in building resilience and sharing best practices that would also benefit the rest of the Philippines and the ASEAN region.

8.4 Concluding Remarks

The Philippines has come a long way in climate change adaptation coming from a disaster-oriented and highly reactive approach to hazards that later shifted to predominantly mitigation responses. Initial strategies on climate change adaptation (CCA) can be considered as "no regret" solutions, or those actions that are beneficial even without climate change. Since then targeted responses are being taken to address the impacts of climate change, particularly in relation to extreme weather events and climate variability. This started the serious mainstreaming of

CCA into policies and development plans, which specifically address the country's and the local governments' vulnerabilities. At the moment, much of the accomplishments of the Philippines in CCA relate to adaptation planning. It is yet to be seen how the adaptation strategies that would be drawn by each local government unit (LGU) would contribute to the ultimate outcome of climate-smart development at different levels of governance in the country. It is therefore hoped that the Climate Change Act would not suffer the same fate of other environmental laws with wanting results.

Caution should be applied in identifying and implementing CCA and the measures implemented do not always lead to resilience. Trade-offs among adaptation strategies should be considered, and should not render any system (or any LGU in particular) more vulnerable. Similar to development initiatives, fragmentation is a major challenge. Hence, not only an action-oriented approach to adaptation is demanded, but also a systems-oriented one to ensure that nothing is left out. This likewise entails the cooperation of different stakeholders in a local government, inter-LGU collaboration (particularly those located in a single watershed), and consistency in governance across scales in working toward achieving a climate-risk resilient Philippines. The proposed local level pilot is the municipality of San Vicente, Palawan located on the northwestern side of the Palawan Province. San Vicente is a demonstration site for CCC's Eco-town framework. The proposed pilot will apply the concepts of multifunctionality, transdisciplinarity, participation, complexity, and sustainability for enhancing climate resilience, and go beyond the administrative border to take into account its ecosystems boundary, to capture systems that influence, and are influenced by the municipality.

Acknowledgments The authors acknowledge the ASEAN-India Green Fund for supporting the ASEAN-India Project on "Enhancing Local Level Climate Change Adaptation in Southeast Asia" (Phase 1). This chapter draws partly on findings of the ASEAN-India Project, specifically the local level needs assessment conducted in 2015.

References

ACCBio (2009) An institutional collaboration for the formulation of the Philippine strategy for climate change adaptation. Department of environment and natural resources and Deutsche Gesellschaft für technische Zusammenarbeit (GTZ). Makati City, Philippines, p 24

Adaptation Knowledge Platform (2012) Scoping assessment on climate change adaptation in the Philippines. Regional climate change adaptation knowledge platform for Asia. Bangkok Thailand, p 46

Balangue TO (2013) Compendium of climate change vulnerability and impact assessment tools. Climate Change Commission, Manila, Philippines

Barletti JPS, Larson AM, Hewlett C, Delgado D (2020) Designing for engagement: a realist synthesis review of how context affects the outcomes of multi-stakeholder forums on land use and/or land-use change. World Dev 127(2020):1–19. https://doi.org/10.1016/j.worlddev.2019.104753

Climate Change Commission (2010) National framework strategy on climate change 2010–2022. Office of the President of the Philippines, Malacanang, Philippines. http://climate.gov.ph/
Climate Change Commission (2011) National climate change action plan 2011–2028. Climate Change Commission, Malacanang Compound, Manila, Philippines
Climate Change Commission (2013) Philippines research and development agenda in support of the National climate change action plan strategic priorities (2011–2028). Climate Change Commission, Malacanang Compound, Manila, Philippines
CCC and GGGI (2014) Demonstration of the eco-town framework project in San Vicente, Palawan Philippines. Climate Change Commission, Manila, Philippines, p 239
De Leon EG, Pittock J (2016) Integrating climate change adaptation and climate-related disaster risk reduction policy in developing countries: A case study in the Philippines. Clim Dev 5:171–184. https://doi.org/10.1080/17565529.2016.1174659
DENR (2012) Climate change adaptation: best practices in the Philippines. Department of environment and natural resources. Manila, Philippines
DOH and WHO (2012) Policies handbook on climate change and health. Department of Health, Manila, Philippines
Doberstein B, Tadgell A, Rutledge A (2020) Managed retreat for climate change adaptation in coastal megacities: a comparison of policy and practice in Manila and Vancouver. J Environ Manage 253:1–10. https://doi.org/10.1016/j.jenvman.2019.109753
Espinas A (2012) Geography and public planning: albay and disaster risk management. Human development network (HDN) Discussion paper series. PHDR Issue 2012/2013 no. 4. HDN. School of Economics, University of the Philippines Diliman, Quezon City, Philippines
Freeman OE, Duguma LA, Minang PA (2015) Operationalizing the integrated landscape approach in practice. Ecol Soc 20(1):24. https://doi.org/10.5751/ES-07175-200124
Finkel T, Duhaylungsod N (2010) Climate change and private sector development integrating green growth strategies into the MSME Development Plan 2010–2016. Promotion of green economic development ProGED project, p 1–81
Gallopin GC (2006) Linkages between vulnerability, resilience and adaptive capacity. Glob Environ Change 16:293–303. https://doi.org/10.1016/j.gloenvcha.2006.02.004
Guingona III T, Bizaon R (2009) Philippine disaster risk reduction and management act of 2010.
Ionescu C, Klein RJT, Hinkel J, Kumar KSK, Klein R (2009) Towards a formal frameworks of vulnerability to climate change. Environ Model Assess 14:1–16
ISET (2013) Climate resilience framework: putting resilience into practice. Institute for social and environmental transition-international, Colorado, USA, p 22
IPCC (2001) Climate change 2001: impacts, vulnerability and adaptation. Contribution of working group 2 to the third assessment report of the intergovernmental panel on climate change. Cambridge University Press, Cambridge, United Kingdom
IPCC (2012) Summary for policy makers. In: Field CB, Barros V, Stocker TF, Dahe, Q, Dokken DJ, Elbi KL, Mastrandrea MD, Mach KJ, Plattner GK, Allen SK, Tignor M, Midgley PM (eds) Managing the risks of extreme events and disasters to advance climate change adaptation. A special report of working groups I and II of the intergovernmental panel on climate change. Cambridge University Press, Cambridge, UK, and New York, NY, USA, p 3–21
IPCC (2007a) Working group II Appendix 1: glossary. Intergovernmental panel on climate change. Geneva, Switzerland, p 104
IPCC (2007b) Summary for policymakers. In: Solomon S, Qin D, Manning M, Chen Z, Marquis M, Averyt KB, Tignor M, Miller HL (eds) Climate change 2007: the physical basis. Contribution of working group i to the fourth assessment report of the intergovernmental panel on climate change. Cambridge University Press, Cambridge, United Kingdom and New York, NY, USA
IPCC (2014) Climate change 2014: impacts, adaptation, and vulnerability—summary for policymakers. In: Field CB, Barros VR, Dokken DJ, Mach KJ, Mastrandrea MD, Billir TE, Chatterjee M, Ebi KL, Estrada YO, Genova RC, Girma B, Kissel ES, Levy AN, MacCracket, S, Mastrandrea PR, White LL (eds) Contribution of working group II to the fifth assessment report

of the intergovernmental panel on climate change. World Meteorological Organization, Geneva, Switzerland, p 34

Kreft S, Eckstein D, Junghans L, Kerestan C, Hagen U (2014) Global climate risk index 2015. Briefing paper. Bonn and Berlyn, Germanwatch, Germnay, p 1–32

Lasco RD, Cruz RVO, Pulhin JM, Pulhin FB (2010) Assessing climate change impacts, vulnerability and adaptation: the case of Pantabangan-Carranglan Watershed. University of the Philippines Los Baños, World Agroforestry Centre and College of Forestry and Natural Resources, p 95

Lasco RD, Pulhin FB, Jaranilla-Sanchez PA, Garcia K, Gerpacio R (2008a) Mainstreaming climate change in the Philippines. Working paper no 62. World Agroforestry Centre (ICRAF), College, Laguna, Philippines, p 23

Lasco RD, Rangasa M, Pulhin FB, Delfino RJ (2008b) The role of local government units in mainstreaming climate change adaptation in the Philippines. World agroforestry centre (ICRAF), Laguna, Philippines and center for initiatives and research on climate adaptation (CIRCA). Albay, Philippines

McCarthy J, Canziani O, Leary N, Dokken D, White K (2001) Climate change 2001: Impacts, adaptation and vulnerability. Contribution of working group II to the third assessment report of the intergovernmental panel on climate change. Cambridge, UK: Cambridge University Press

Mendoza MET (2014) Methodology for a capital-based composite index for climate change vulnerability assessment: the ENS 299 Seminar Cooling the Hotspot, Philippine climate change initiatives: where do we go from here?. CFNR Auditorium, UPLB, 13 Nov 2014

Nelson DR, Adger WN, Brown K (2007) Adaptation to environmental change: Contributions of a resilience framework. Annu Rev Environ Resour 32:395–419

O'Brien K, Hochachka G (2010) Integral adaptation to climate change. J Integral Theory and Pract 5(1):89–102

PGA & CIRCA (Provincial Government of Albay and Center for Initiatives and Research on Climate Adaptation (2010) Province of Albay, Philippines: responding to the challenge of disaster risk reduction and climate change adaptation. Strengthening climate resilience (scr) case study summary. Philippines. http://www.preventionweb.net/files/section/230_Philippinesalbaycasestudy.pdf. Accessed 25 April 2016

Republic of the Philippines (2014) The Philippines' second national communication to the United Nations framework convention on climate change, p 110

Sanchez PB (2015) The national climate change action plan of the Philippines. https://www.slideshare.net/NAP_Global_Network/philippinesnationalclimatechangeactionplan. Accessed 28 Jan 2016

Sandoval Jr R, Baas S (2013) Adapting to climate change: the cordillera experience. Knowledge center on climate change 3(2). SEAMEO-SEARCA, College, Los Banos, Laguna. http://www.climatechange.searca.org/images/stories/adncca/sandoval-and-baas-ADN.pdf. Accessed 25 April 2016

Smit B, Wandel J (2006) Adaptation, adaptive capacity and vulnerability. Glob Environ Change 16:282–292. https://doi.org/10.1016/j.gloenvcha.2007.05.002.Stern

Stern Report (2007) The economics of climate change: the stern review. Cambridge University Press. http://www.hm-treasury.gov.uk/sternreview_index.htm. Accessed 17 Nov 2010

Tan, Alan KJ (2013) Preliminary assessment of Philippine's environmental law, APCEL Report: Philippines. http://sunsite.nus.edu.sg/apcel/dbase/filipino/reports.html. Accessed 13 July 2016

Tapia MA, Pulhin JM, Peras RJJ (2014) Vulnerability and adaptation to climate change of selected community-based forest management areas in Oas, Albay Philippines. Asia Life Sci 23(2):567–592

Toda L, Orduña JR, Lasco RD, Santos ST (2015) Assessing social vulnerability to climate-related hazards among Haiyan-affected areas in Leyte, Philippines. Clim Disaster Dev J 1(1): 42–57. https://doi.org/10.18783/cddj.v001.i01.a05

UNISDR (2009) Adaptation to climate change by reducing disaster risks: country practices and lessons. Briefing Note 02. United Nations international strategy for disaster reduction. Geneva

Uy N, Takeuchi Y, Shaw J (2011) Local adaptation for livelihood resilience in Albay, Philippines. Environ Hazard 10(2):139–153. https://doi.org/10.1080/17477891.2011.579338

World Bank (2010) A strategic approach to climate change in the Philippines. In: Ancheta C, Bojo J, Dato V, Heister J, Kariuki M, Morton J, Trohanis Z, Tuyor J, Villaluz M, Virtucio f, Wedderburn S, Zhang Y (eds) Sustainable development department, East Asia & Pacific Region, World Bank. http://siteresources.worldbank.org/INTPHILIPPINES/Resources/PH_CC_Strategy_April_19_2010.pdf. Accessed 19 April 2020

Chapter 9
Climate Change Adaptation in Singapore

Ankit Joshi and Andreas Schaffer

Abstract The Nationally Determined Contribution (NDC) has stated that Singapore has pledged to reduce its emission levels by 36% in 2030. This is complemented by actions to adapt adequately to climate change as Singapore is one of the most densely populated countries. This chapter highlights the early warning measures that are being put into place to minimize adaptation costs for this island state. In an effort to cultivate its own body of knowledge and expertise in climate science and climate modelling, the Meteorological Service Singapore set up the Centre for Climate Change Research Singapore to study the impacts climate change on the island's biodiversity and greenery, its network and building infrastructure, and more importantly, its water resources and drainage. Two pilots are proposed that would benefit both the island state and ASEAN Member States. The first is on the feasibility and opportunities for mobilizing private sector finance options to advance urban infrastructure adaptation. The second is on developing a holistic approach for assessing food security challenges and response options using scenario development, downscaled climate projections, and modelling of major food crop markets and commodity prices.

Keywords Adaptation plan · Water resources · Sea level · Food security Singapore

9.1 Introduction

Singapore is a dynamic, multiracial, and multicultural society also known as Asia's 'garden city'. The population consists of Chinese (74.1%), Malay (13.4%), Indians (9.2%), and other races. This diversity also translates into a unique mix of religions, languages, cultural traditions, and cuisines. It is a developed and successful free-market economy with a per capita GDP higher than that of most developed

A. Joshi (✉) · A. Schaffer
Independent Advisor, Asian Disaster Preparedness Center, Bangkok, Thailand
e-mail: ankitjoshi11@gmail.com

A. Schaffer
e-mail: a.schaffer2017@gmail.com

J. J. Pereira et al. (eds.), *Climate Change Adaptation in Southeast Asia*,
Disaster Risk Reduction, https://doi.org/10.1007/978-981-16-6088-7_9

countries. In 2019, the country's GDP was USD 372 billion and its per capita GDP was USD 65,233.3 (World Bank 2020). The economy depends mainly on exports, particularly of consumer electronics, information technology products, medical and optical devices, pharmaceuticals, and on its vibrant transportation, business, and financial services sectors. Decoupling economic growth and environmental degradation is a major challenge for Singapore. Efforts are underway to understand the dynamics between environmental policy, capital flow, technological innovation, and economic development (Chen and Taylor 2020).

9.1.1 Context

Singapore is naturally vulnerable to climate change because of its small size and low-lying topography. Most of Singapore's land area of 716 km^2 is less than 15 m above sea level. With a population of approximately 5.6 million (as of 2017), Singapore is one of the most densely populated countries in the world with a population density of about 7,796 people per km^2. Therefore, Singapore must adapt adequately to climate change.

In 2012, the Government of Singapore released the National Climate Change Strategy document outlining the city-state's plans to address climate change adopting a whole-of-nation approach. This document contains key elements of Singapore Government's climate strategy including reducing emissions across various sectors, building adaptive capabilities, capitalizing on green growth opportunities, and forging partnerships on climate change action.

In 2016, the Singapore Government formalized its Nationally Determined Contribution (NDC) under the United Nations Framework Convention on Climate Change (UNFCCC), which was updated in 2020. The nation state has pledged to reduce its emissions intensity by 36% from 2005 levels by 2030, and to stabilize its emissions with the aim of peaking around 2030 (Government of Singapore 2020).

9.1.2 Climate

Singapore has witnessed nine of the 10 warmest years since 1998 and El Niño conditions prevailed in eight of these. Changes have been recorded in the frequency of extreme high and low temperatures across the island since the 1970s (MSS 2015). From 1972 onwards, Singapore has witnessed an upward trend in the number of warm days and warm nights along with a decrease in the number of cool nights.

Rainfall records have been well maintained by the Meteorological Service Singapore (MSS) since 1869. Singapore has seen noticeable annual variability in rainfall. This variability is influenced greatly by the El Niño Southern Oscillation (ENSO) during the southwest monsoon and the inter-monsoon season that follows. The southwest monsoon season has been experiencing more rainfall during stronger El Niño events. In 1997, Singapore experienced its worst El Niño event; during the period of

June–September of this year, there was a 53% drop in rainfall. However, it cannot be said that there is direct link between strong El Niño events and local rainfall. Less intense El Niño events in some years, like that in 1963, have had similar effects on rainfall too.

Recent findings reveal that the total annual wet-day rainfall as well as the frequency and intensity of precipitation extremes have increased significantly over the past three decades (Li et al. 2018). The global warming signal is masked by local effects in the urban city, having implications on climate change adaptation.

Singapore is also affected by heat stress, which is the discomfort experienced during very hot days; this is measured as the 'wet-bulb globe temperature' (WBGT) and 'apparent temperature'. This empirical measurement records a person's comfort level with regard to temperature, humidity, wind, and sunlight. Since 1995, there have been 10 years when heat stress or WBT figures have been significantly higher. The El Niño and La Niña events have an effect on Singapore's annual average WBT. In 1998, the strongest El Niño event on record (MSS 2015) happened, and it saw a peak in WBT; the La Niña event occurred only once in that period.

In recent years, Singapore has seen bouts of high temperatures and very intense thunderstorms that have led to flash floods. The annual mean temperature rose from 26.6 °C in 1972 to 28.3 °C in 2015, which was both the warmest year and the second-driest year ever recorded in the island state. In early 2014, Singapore also experienced its longest dry spell since records began in 1869. It is projected that Singapore temperatures could rise by between 1.4 °C and 4.6 °C by the end of this century (2070–2099), while mean sea levels could increase by between 0.25 m and 0.76 m in the same period (MSS 2015).

9.2 National Perspective

9.2.1 National Arrangements and Institutions

Singapore has taken a proactive and long-term government-lead approach to tackling climate change. Its plans have continuously been recalibrated in accordance with the latest scientific findings in climate change science. At the country level, stakeholder consultations are held as and when required involving the scientific community under the guidance of the respective ministries and government departments. Singapore's approach to climate change adaptation, given the constraints and unique circumstances in which it operates, must be looked upon as a special case in the climate change adaptation landscape.

In Singapore, the Inter-Ministerial Committee on Climate Change (IMCCC) enhances Whole-of-Government coordination on climate change policies to ensure that Singapore is prepared for the impacts of climate change. Established in 2007, IMCCC is chaired by the Deputy Prime Minister; it includes the Minister for the Environment and Water Resources, the Minister of Finance, the Minister for Foreign

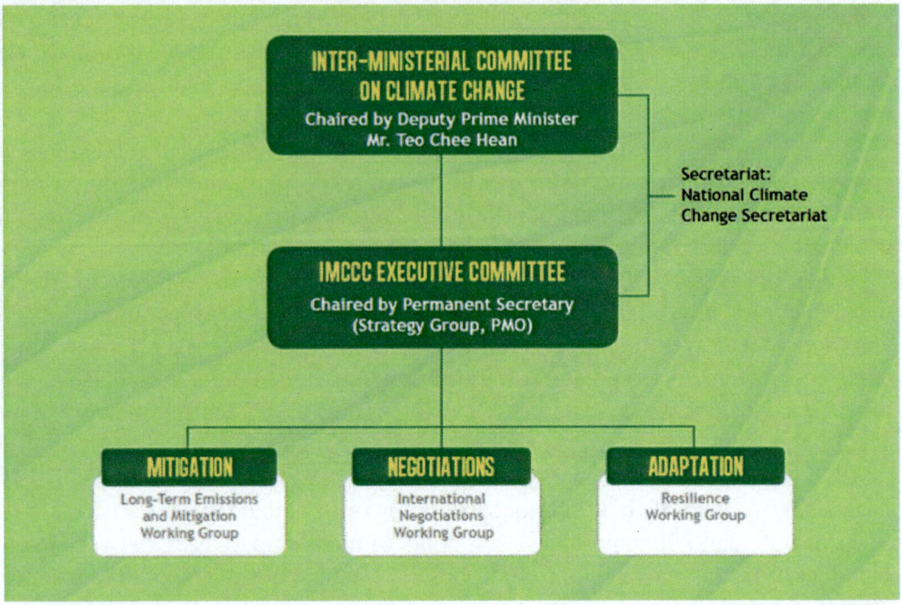

Fig. 9.1 The framework of the inter-ministerial committee on climate change. *Source* NEA (2018)

Affairs, the Minister for National Development, the Minister for Trade and Industry, and the Minister for Transport (Fig. 9.1).

There are three working groups under IMCCC: The International Negotiations Working Group develops Singapore's international climate change negotiations strategy under the UNFCCC; the Long-Term Emissions and Mitigation Working Group studies how Singapore can stabilize its long-term emissions; and the Resilience Working Group evaluates Singapore's vulnerability to the effects of climate change and recommends long-term plans to build Singapore's resilience. These three committees report to an Executive Committee. In July 2010, The National Climate Change Secretariat, which is under the Prime Minister's Office, was formed to ensure effective coordination of Singapore's domestic and international policies, plans, and actions on climate change.

The National Climate Change Strategy 2012 provides a comprehensive overview of Singapore's response to climate change (NCCS 2012). It reaffirms the country's pledge to cut down emissions by 16% below 2020 business-as-usual levels (BAU) if there is a legally binding global agreement. Steps have already been taken by the country to bring down the levels of emissions; the target for this 7–11% below 2020 BAU levels.

Climate Change Adaptation is a continuous process and the government views it as a long-term engagement going into the future. The National Climate Change Strategy of 2012 highlights six sectors particularly vulnerable to climate change and proposes appropriate adaptation strategies. These initiatives are usually based on an appraisal of the latest available scientific research and are designed by subject matter experts under the guidance of the various government agencies. Coastal protection, addressing flood risks, management of water resources, safeguarding biodiversity and greenery, strengthening resilience in public health, and creating a cooler built environment continue to be priority areas for the government.

The Sustainable Singapore Blueprint 2015 updates and extends the first blueprint published in 2009. Based on a consultation process involving over 6,000 members of the public, the new version is published by the Ministry of the Environment and Water Resources, and the Ministry of National Development (Government of Singapore 2014). Targets include 200 ha of sky rise greenery by 2030 from 61 ha in 2013; 140 L domestic water consumption per capita per day by 2030 from 150 L at present; reduce flood prone areas to 23 ha by 2030 from 34 ha at present; and park provision of 0.8 ha/ 1000 population by 2030.

9.2.2 Resilience and Coordinated Research

Singapore is trying to address climate change early to get the most benefits while incurring minimal adaptation costs (NCCS 2012). Flexible adaptation plans will allow for the addition of the latest in climate data and scientific developments. A multi-agency Resilience Working Group (RWG), led by the Ministry of the Environment and Water Resources (MEWR) and the Ministry of National Development (MND) under the IMCCC, will be responsible for assessing how vulnerable Singapore is to climate change up to the year 2100.

The agencies that make up the RWG are the Ministry of Health (MOH), the Ministry of Finance (MOF), the Agri-Food and Veterinary Authority (AVA), the Building and Construction Authority (BCA), the Civil Aviation Authority of Singapore (CAAS), the Energy Market Authority (EMA), the Housing and Development Board (HDB), the Info-communications Media Development Authority (IMDA), JTC Corporation, the Land Transport Authority (LTA), the Maritime and Port Authority of Singapore (MPA), the National Climate Change Secretariat (NCCS), the National Environment Agency (NEA), the National Parks Board (NParks), PUB, Singapore's National Water Agency, the Singapore Land Authority (SLA), and the Urban Redevelopment Authority (URA).

The Government of Singapore has acknowledged the importance of science to support adaptation policies and actions (NCCS 2015). Hence, considerable resources have been allocated towards enhancing their knowledge of the various aspects of climate change-related impacts and vulnerabilities. The government is building up its expertise in climate and weather sciences within the MSS and local research institutions as well as forming partnerships with international experts.

In an effort to cultivate its own body of knowledge and expertise in climate science and climate modelling, the Meteorological Service Singapore set up the Centre for Climate Change Research Singapore (CCRS) (Fig. 9.2). The CCRS is steered by the International Scientific Advisory Panel on Climate Science and relevant experts and institutions through a Climate Science Experts Network. This Network draws upon

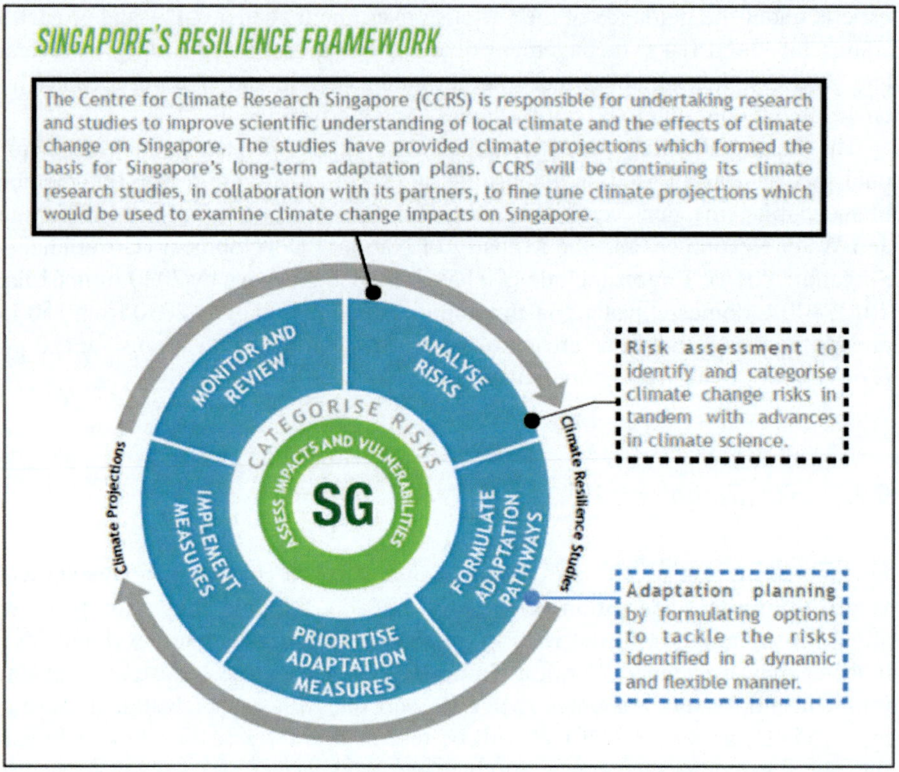

Fig. 9.2 The framework for building climate resilience in Singapore. *Source* (NEA 2018)

the expertise within local academia to complement the expertise of CCRS. This has not only helped build up Singapore's expertise in climate science, but also informs national preparedness plans for climate change and crucial policy-level decisions.

Additionally, local research institutes in Singapore have been given a mandate to conduct interdisciplinary and translational research on climate change and coastal protection. The National University of Singapore (NUS) and Nanyang Technological University (NTU) have set up several research institutes:

- The Singapore-Delft Water Alliance (SDWA): SDWA promotes understanding of the water cycle using environmental sensing methods, intelligent data processing, and modelling approaches. This includes expertise on real-time forecasting and warning, including analysis and simulation of extreme events, for informing decision support systems.
- Tropical Marine Science Institute (TMSI): TMSI is a Centre of excellence for research, development, and consultancy in tropical marine science as well as environmental science. With its multidisciplinary research laboratories and active international links, it handles projects relevant to physical oceanography, acoustics, marine biology, marine mammals, biofuels, water resources, and climate

change. TMSI was involved in Phase 1 of the Climate Change Study and is helping with the current Risk Map Study for Singapore.

- Earth Observatory of Singapore (EOS): EOS is a Research Centre of Excellence established to conduct fundamental research on earthquakes, tsunamis, volcanic eruptions, and climate change in Southeast Asia and elsewhere. EOS's research focuses on regional and local consequences of global climate change, by investigating drivers of global climate change in the tropics, including the Western Pacific Warm Pool and the Indian Ocean Dipole.
- Institute of Catastrophe Risk Management (ICRM): ICRM is the first institute of its kind in Asia working on multidisciplinary risk and management research. There are only a few such centres in the world that focus on catastrophe-triggered insurance and reinsurance risks, societal risks, sovereign risks including some non-traditional risks.
- Maritime Research Centre (MRC): MRC conducts maritime research and development (R&D) on marine and offshore engineering, hydrodynamics, shipping, business, and logistics in collaboration with local and international organizations. MRC's work includes beach stabilization research, and the role of mangroves in protecting mudflats from erosion.
- NTU-JTC Industrial Infrastructure Innovation (NTU-JTC I3) Centre: NTU-JTC I3 works around four themes, namely reclamation and marine infrastructure, sustainable infrastructure, underground infrastructure, and infrastructure systems and materials. It also aims to build new capabilities and to scale up technical leadership competencies in the area of industrial infrastructure development in order to promote the development of economically feasible and sustainable industrial infrastructure solutions in Singapore.

Since 2013, the Singapore government has started a 5-year project to take stock and monitor the entire island's carbon emissions to understand the role of the country's greenery in absorbing these emissions. The intention is to develop a monitoring system that tracks how much trees, soil, and possibly even the grass help to reduce greenhouse gases. An accurate inventory is needed since international groups, using different calculating techniques, have come up with widely fluctuating emissions figures. In order to do this, the National Parks Board has roped in the National Institute of Education (NIE), the Austrian Natural Resources Management, and International Cooperation Agency (Anrica) for technical collaboration.

Additionally, a modified Urban Ecosystem Services Review (ESR) was conducted for Singapore based on World Resources Institute's ESR (Sieber et al. 2015). This involved an assessment of the air quality regulation service by Singapore's greenery using carbon storage analysis. The assessment was based on the InVEST model including current land cover data, carbon pools, and a valuation model with a value of carbon (USD/metric ton), the market discount, and the annual rate of change in price. The findings revealed that areas with the highest amount of stored carbon are forested, i.e. national park areas such as MacRitchie Reservoir, the Bukit Timah Nature Reserve, the wetland reserves, and green areas around the airport. Other smaller parks and community gardens, not reflected in the results of this study are also believed to store carbon and help in keeping the urban environment clean and cool.

9.2.3 Vulnerable Sectors and Regions

The Intergovernmental Panel on Climate Change (IPCC) reports that global climate is undergoing drastic change (IPCC 2014). Global temperatures are projected to rise by 0.3–4.8 °C by 2100 as above present levels, while global mean sea levels are expected to increase by 0.26–0.82 m over the same period. This assessment is based on future scenarios representative of greenhouse gas concentrations. Extreme precipitation events over the region are also projected to be more intense and frequent.

To translate the implications of IPCC's findings for Singapore and to prepare and plan adaptation actions, Singapore is presently conducting the second National Climate Change Study (NCCS 2015). On April 2015, the Meteorological Service Singapore's (MSS) Centre for Climate Change Research Singapore (CCRS) released the findings of Phase 1 of this Study. The study uses a subset of models for two future trajectories of greenhouse gas concentrations in the atmosphere, which were downscaled to a higher resolution of 12 km for the Southeast Asian region. The first trajectory (RCP 8.5) is a 'business as usual' scenario with continuous use of fossil fuels with no mitigation efforts and thus results in very high greenhouse gas emissions that continue to increase throughout the twenty-first century. The second trajectory (RCP 4.5) assumes future action to control emissions, which peak in mid-century and then decline. The results from the downscaled projections of these two scenarios are the basis for the study's findings. The outcomes of this study are consistent with those of the IPCC 5th Assessment Report (AR5), which projects higher temperatures and more extreme rainfall for Singapore and the surrounding region along with global sea level rise.

For the period of 2070–2099 (relative to 1980–2009), average surface temperatures across Singapore will increase by 1.4–2.7 °C under the RCP 4.5 greenhouse gas trajectory and by 2.9–4.6 °C for RCP 8.5 (NCCS 2015). These findings imply that the unusually warm temperatures Singapore encounters occasionally could become the norm in the future and days with 'record' temperatures above those experienced historically will become more frequent. Due to the high levels of humidity in Singapore, these projected rises in temperature will lead to increasing thermal discomfort and heat stress for those working outdoors.

For rainfall, the projections from most models show an increasing trend, particularly during the wetter season, in the intensity and frequency of heavy rainfall events over Singapore. For seasonal average rainfall, the contrast between the wetter and drier months is projected to become more pronounced, especially for the RCP 8.5 trajectory and by the end of the century. The study projects an upward trend in seasonal mean rainfall during the wet season of November to January, as well as greater dryness during months that are already relatively drier in the current climate (February, and June to September).

As far as sea level rise is concerned, for the period of 2070–2099 (relative to 1980–2009), the study projects a mean sea level rise of 0.25–0.60 m under the RCP 4.5 greenhouse gas trajectory and 0.35–0.76 m for RCP 8.5 (NCCS 2015). Under the 'high-emissions' RCP 8.5 greenhouse gas trajectory, the study projects a total rise of 0.45–1.02 m by the year 2100. The study projects on consistent future changes in the amplitude of extreme sea level events, such as wind driven coastal storm surges and high waves. These events will, however, occur against a backdrop of increased mean sea levels.

The next phase will build on the projections from Phase 1 to examine the impacts of climate change on areas such as water resources and drainage, biodiversity and greenery, network infrastructure, and building infrastructure. This in turn will guide Government agencies in their planning and will serve as an input in the development of climate change resilience plans for the island state.

9.2.4 Adaptation Actions

The Singapore Government has taken substantial adaptation actions in the past to protect the island state from the impacts of climate change based on solid climate science. The government views climate change adaptation as a continuous process and is proactively aiming to integrate the latest findings from Phase 2 of the 2nd National Climate Change Study, once they are released. Adaptation actions by the government have so far revolved along the following: coastal protection, addressing flood risks, management of water resources, safeguarding biodiversity and greenery, strengthening resilience in public health and food supply, creating a cooler built environment, maintaining essential services, and keeping buildings and infrastructure safe.

Coastal Protection: Singapore being a low-lying island is threatened most by sea level rise. Majority of Singapore lies within 15 m above mean sea level and about 30% of Singapore landmass is less than 5 m above the mean sea level. Currently, approximately 70–80% of Singapore's coastal areas are protected with stone embankments or hard walls, which help safeguard against coastal erosion. The remaining coastal areas comprise natural areas such as beaches and mangroves. The country has tried to address these challenges over the last decade. Since 1991, in accordance with regulations, all land reclamation projects have been built at least 1.25 m above the highest recorded tide level. From 2011 onwards, the minimum requirement has been set at four metres above the mean sea level, one metre more than before; this is to protect the land from long-term sea level rise.

Related fields such as coastal engineering, wave dynamics, coastal morphology, and hydrodynamics come into play in the protection of coastal areas. Several public agencies, including the Building and Construction Authority, Housing & Development Board (HDB), and JTC Corporation (JTC) have developed capabilities in these areas; they are actively contributing in coastal engineering and coastal protection works (BCA 2013). A Coastal and Project Management Department has been established in the Building and Construction Authority (BCA) to focus on coastal protection and adaptation issues and initiated a Risk Map Study (RMS) to identify specific coastal areas that are at risk of inundation and the potential expected damages. When completed towards the end of 2013, this study covered the entire coastline of Singapore with several topographical surveys. The results from this RMS will complement the Coastal Adaptation Study (CAS) that started in October 2013, with a mandate to develop long-term coastal protection strategies for Singapore. There has also been some interest in 'soft' coastal protection practices, which include the use of plants (such as mangroves and sea grass) for enhancing coastal protection along with beach and dune nourishment. Local institutions have also been involved in research to improve protection measures in these areas as well as in developing new approaches. One such project involved the restoration of mangroves on the northeastern coast of Pulau Tekong using a combination of mangrove plants, rocks, and biodegradable rings. Besides providing coastal protection and restoration habitat, this also enables the integration of climate change adaptation and disaster risk reduction. As a result, Pulau Tekong has now become a test bed for sustainable soft engineering solutions.

Addressing Flood Risks: The maximum hourly rainfall over Singapore has increased from an average of 96 mm in 1980 to 117 mm in 2012. Moreover, the 2nd National Climate Change Study for Singapore predicts an increase in the frequency and intensity of heavy rainfall events in the future and more so over the wetter seasons. In view of greater weather uncertainties and urbanization, the PUB has revised its storm water management procedures to improve flood protection for Singapore. PUB's new strategy aims to optimize the management of storm water using a holistic source-pathway-receptor approach. This approach addresses catchment-wide solutions to achieve higher drainage and flood protection standards. It covers the entire drainage system by addressing not just the pathway over which the storm water flows (i.e. 'Pathway' solutions such as drain widening and deepening), but also controlling rainwater at source (i.e. 'Source' solutions such as on-site detention) and the areas where floods are likely to occur (i.e. 'Receptor' solutions such as specification of platform levels).

Measures have also been taken to reduce flood risks. Over the last 30 years, flood prone areas have been reduced from about 32 km^2 in the 1970s to about 0.49 km^2 as at January 2012. In accordance with PUB's Code of Practice for Surface Water Drainage (COP), minimum platform level for developments has been raised to further safeguard against the longer term effects of sea level rise and increase in rainfall intensity. In 2011, PUB also raised design standards for new drains to cater for more intense extreme rainfall events. Depending on the size of the catchment, this translates into 15–50% increase in drainage capacity. From 1 January 2014 onwards, developers/owners of land size 0.2 hectares or more are required under PUB's COP to implement measures to slow down surface runoff and reduce the peak flow of storm water into the public drainage system by implementing on-site detention measures such as green roofs, rain gardens, and detention tanks. In addition, PUB is studying the feasibility of innovative underground drainage and reservoir systems, where tunnels will bring excess storm water into underground caverns for storage. This system has the potential to reduce flood risks significantly.

Management of Water Resources: Increased rainfall intensity and weather variability present grave challenges for water resource management in Singapore. Spells of drought can affect the reliability of Singapore's water supply, while sudden bursts of intense rainfall can overwhelm the drainage system and lead to flash floods. These extreme weather events are likely to increase in intensity and frequency with climate change. In order to tackle this challenge, PUB has implemented a robust and diversified water supply system through the Four National Taps (local catchment water, imported water from neighbouring Malaysia, NEWater, i.e. water that has undergone a stringent treatment and purification process, and desalinated water) to ensure a sustainable supply of water for Singapore's consumers. NEWater and desalinated water supplies do not depend on rainfall and therefore, can be used to supplement water from local reservoirs as needed. PUB aims to complete the construction of its fourth desalination plant in the coming years. PUB is also looking to build a fifth desalination plant on Jurong Island. It is expected that by 2030, NEWater and desalinated supply would meet approximately 80% of Singapore's water demand. Water conservation is an integral and complementary strategy to ensure Singapore's long-term water sustainability. Through public outreach and other measures, the daily per capita domestic water consumption in Singapore has dropped from 165 L in 2003 to 153 L in 2011. The long-term target for daily per capita domestic consumption of water set out in the Sustainable Singapore Blueprint is 147 L in 2020.

Safeguarding Biodiversity and Greenery: Fragmentation of forests in Singapore makes them vulnerable to long-term changes like increased likelihood or duration of drought and higher average temperatures. Consequently, roadside trees could be uprooted in storms and become a safety hazard. Wetlands may be affected by rainfall changes, sea level rise, or water quality changes related to warming and changes in precipitation. Sea level rise, in particular, will be a challenge for mangroves, which cannot retreat inland. Corals, which require sunlight, may not be able to grow upwards quickly enough to keep pace with rising sea levels. In addition, a 1–2 °C rise in seawater temperatures is expected to cause widespread coral bleaching. Adverse effects on coastal vegetation could lead to changes in rate of soil erosion along the shore. The mitigating effects of coastal vegetation on flooding or tidal surges could also be adversely impacted. Additionally, increased rainfall due to climate change may affect slope stability by adversely affecting the vegetation, which normally holds soil in place on slopes. To safeguard the biodiversity of Singapore against the potential impacts of climate change, NParks is working with all concerned stakeholders to protect existing species at high risk of being impacted, increasing connectivity of various green spaces across Singapore, and enhancing the resilience of ecosystems (Government of Singapore 2010). This includes measures to restore forest and mangrove areas, diversify plant species, intensify tree planting and increase connectivity between green areas. NParks established Singapore's first marine park at Sister's Islands in 2014 to protect Singapore's marine biodiversity (Government of Singapore 2010). The Singapore Civil Defence Force (SCDF) also plans to step up patrols at fire hot spot areas for speedy response. Additionally, SCDF, NParks, and the Met Service plan to develop a Fire Probability Index to indicate the risk of bush fires for different areas. In order to keep the city green, tree management and maintenance is also being enhanced. Going forward, NParks will continue to lead the planning and maintaining parks and green spaces island-wide as part of the national effort to lower ambient temperatures.

Strengthening Resilience in Public Health and Food Supply: Due to its tropical location, vector-borne diseases, such as dengue are endemic to Singapore. Changes in the weather pattern, such as temperature and rainfall increase, could create ideal conditions for mosquitoes to breed and viruses to replicate faster, leading to an increase in vector-borne diseases. Similar trends were observed in late 2015, when there was a spike in dengue cases partly due to weather changes caused by the El Niño.

A nation-wide integrated programme, which combines mosquito, virus, and human surveillance as well as public education and participation, law enforcement and research has been implemented to minimize dengue cases in Singapore (NEA 2018). The feasibility of using the male Wolbachia-carrying *Aedes aegypti* mosquito to help suppress the *Aedes aegypti* mosquito population is also being explored. *Aedes aegypti* mosquitoes are dengue vectors. There is empirical evidence to show that when male Wolbachia-*Aedes aegypti* mosquitoes mate with female *Aedes aegypti* mosquitoes without Wolbachia, the females produce eggs that do not hatch, thereby reducing the *Aedes aegypti* population over time. A study on the relationship between climatic factors—such as temperature, humidity, and rainfall—and public health risks such as dengue fever, heat disorders, and respiratory diseases is also ongoing (NEA 2018). The study also looks into forecasting the risk of disease transmission in different scenarios. NEA's Environmental Health Institute (EHI), a World Health Organization Collaborating Centre for the Reference and Research of Arbovirus and their Associated Vectors, will leverage its research capability in vector-borne diseases to support efforts to strengthen Singapore's public health resilience to climate change. Rising temperatures and reduced rainfall affect the severity of haze events. Smoke haze from land and forest fires in neighbouring countries is carried to Singapore by the prevailing winds. This not only creates health problems in the region but also impacts the economy, tourism and the aviation industry. Singapore views the issue of trans-boundary haze very seriously. Together with the Ministry of Health (MOH), NEA coordinates measures for haze—related illnesses through the National Haze Task Force.

Singapore imports more than 90% of its food. In the future, crop failure and low crop yields due to rising temperatures are expected to impact the supply of items such as rice and vegetables. The fisheries sector is also expected to harvest smaller catches affecting the supply of seafood. In order to minimize the risk of food supply disruptions, the Agri-Food and Veterinary Authority of Singapore (AVA) adopts a diversification strategy. Food is imported from countries in different regions of the world to diversify the food supply (Singapore Food Agency 2006a). This essentially means that if a country or region is impacted adversely by an extreme event, Singapore will be insulated from sudden food shortages and price volatility by relying on the other food sources. In Singapore, AVA is working with local farmers to boost productivity using R&D and technology. Examples include innovative indoor farming and close containment aquaculture. Indoor farming conducted by Panasonic Factory Solutions Asia Pacific uses multi-tier racks with LED lighting to cultivate different vegetables such as mizuna, rocket salad, lettuce, and mini red radish in optimized conditions. Closed Containment Aquaculture Systems (CCAS) such as Recirculating Aquaculture Systems are used to protect fish from adverse environmental conditions. Farms with such systems achieve higher productivity, better water resource efficiency, and space optimization. Singapore Aquaculture Technologies' fish farm produces about 50 tons of seabass a year using CCAS. Seawater, pumped from a depth of four metres, is treated so that the farm is not affected by plankton bloom and other aquatic diseases.

Creating a Cooler Built Environment: Singapore may also experience warmer temperatures as a result of the urban heat island effect which can lead to heat stress particularly among the elderly and the sick. Greater use of air-conditioning will increase Singapore's energy demands. While there may be sufficient generation capacity to meet such an increase, it will eventually result in higher domestic carbon emissions. The Energy Market Authority and Building and Construction Authority are studying Singapore's urban temperature profile and the energy consumption of buildings to understand how temperature increase and wind changes will affect the population at large. The Urban Redevelopment Authority is working with the Housing Development Board (HDB), National University of Singapore (NUS), and Institute of High Performance Computing (IHPC) on a study to understand the effects of the built environment and urban greenery on micro-climatic conditions such as airflow and temperature. The study will also identify hot spots and relatively cooler areas in Singapore, and provide actionable recommendations on the planning and design of public spaces and buildings by modifying building layouts and having larger green spaces. The Ministry of Health (MOH) is working with MSS to develop a heat stress information system for the public to provide information that can help the public plan and manage outdoor activities.

Maintaining Essential Services: Efforts are being taken to ensure that essential services in Singapore are not impacted due to the adverse effects of climate change. Climate change is expected to impact public transport, i.e. the MRT operations by way of flooding during intense rainfall thus impacting the 2.7 million passenger trips taken per day. To protect commuters, the Land Transport Authority (LTA) has installed flood barriers at low-lying MRT stations to prevent water from entering them. In the future, new underground MRT stations will have to meet requirements for additional flood protection. Steps have been taken to ensure air connectivity by minimizing the risk of flooding at Changi airport from rising sea levels and intense rains. The drainage system is being upgraded with water detention tanks and pumps. The future terminal five is to be built 5.5 m above mean sea level. With respect to sea connectivity, the Maritime Port Authority of Singapore (MPA) is monitoring the resilience of the existing and planned ports. The new Tuas terminal has been designed keeping in mind the latest sea level projections and will be built two metres higher than the highest observed water level. The Energy Market Authority is aware of the impacts of floods and increased ambient temperatures on the energy infrastructure of Singapore and is making plans to increase its resilience. Additionally, the Infocomm Development Authority is working with service providers to enhance the resilience of telecommunication networks to climate change.

Keeping Buildings and Infrastructure Safe: Projected changes in temperature, rainfall, and wind speeds are not likely to affect the structural integrity of buildings as long as buildings adhere to the stipulated building codes and are properly maintained. However, BCA and the Housing & Development Board (HDB) are conducting additional assessments to understand the potential impacts of higher temperatures, rainfall, and wind speeds on buildings and building attachments so that they can recommend adaptation measures to enhance their resilience (BCA 2013). Heavier and more frequent rainfall may affect the stability of slopes and trigger slope failures. To address this, BCA is studying appropriate and holistic adaptation strategies to address the increased risk of slope failure due to climate change.

9.2.5 Education and Capacity Building

NEA plays a critical role in promoting climate change awareness and action through education, community and national campaigns, capacity building, and by recognizing the green practices and efforts of the public at large. School going students are introduced to environmental education at school and have climate change as a subject in the general paper along with economics, geography, and the sciences. This enables them to understand how climate change will affect Singapore. Additionally, school visits to power plants, the meteorological station, and green buildings in Singapore are organized to highlight the impacts of climate change and the various solutions available to reduce emissions. Since 2011, NCCS has been organizing a National Climate Change Competition for Schools and tertiary institutions to help raise awareness amongst school going children.

Community level and national level campaigns as the annual flagship Clean & Green Singapore Campaign encourages Singaporean citizens and residents to safeguard the environment by adopting environmentally friendly lifestyles. NEA's 10% Energy Challenge Programme encourages Singaporean citizens and residents to reduce energy consumption and adopt energy-efficient practices. Additionally, energy-labelling schemes enable users to make conscious decisions while buying household appliances such as air conditioners and refrigerators. There is also a fuel economy-labelling scheme that provides information on the efficiency of motor vehicles in the market.

NEA also coordinates several capacity building programmes. These include NEA's Youth Environment Envoy Programme, which prepares students as 'green envoys' to influence their peers and the larger community to be environmentally conscious. Additionally, NEA's 3P Partnership Fund helps organizations in organizing environmental education and outreach activities on climate change and other environment-related topics. Companies are given avenues to participate in NEA's Corporate and School Partnership Programme to help nurture young student leaders through joint environmental programmes and mentorship. BCA's 'Build it Green' (BiG) Club trains youth advocates and promotes greater understanding of green buildings.

Efforts are also made to recognize green practices and actions. One of the most prominent initiatives is the annual President's Award for the Environment—the highest accolade for individuals, organizations, and companies with outstanding contributions to environmental and water sustainability in Singapore. The award serves to demonstrate the Government's support and appreciation for the efforts of the 3P sectors to achieve a sustainable environment. Other awards include NEA's EcoFriend award, which recognizes individuals who have contributed to protecting and improving the environment; BCA's Singapore Green Building Council (BCA-SGBC) Green Building Individual Awards recognize the work of exceptional industry professionals and MPA's Maritime Singapore Green Initiative, which recognizes companies that undertake environmentally friendly shipping practices beyond what is internationally mandated.

In addition to this, the Singapore Environment Institute (SEI), which is the training and knowledge division of NEA, is responsible for the documentation and transfer of knowledge from its experienced staff and other relevant sources to its next generation of officers through training programmes. It also develops and up-skills the local industry's manpower capabilities, thereby adding value to Singapore's environmental arena. SEI's public lectures feature renowned environmental speakers who catalyze thought leadership and foster communities of best practices.

9.3 Adaptation Priorities

9.3.1 Approach

The potential pilot projects on climate change adaptation were evaluated based on several aspects. These include feasibility and opportunities for mobilizing private sector finance options, geographic area and stakeholder profiles, prospective impacts on resources and implications for regional security, constraints and capacity needs as well as potential for institutional arrangements and co-financing. Based on the evaluation, two pilots were identified in order of priority. These are on Climate Change Adaptation Finance and Resilient Food Supply Chains in ASEAN. The proposed pilots need further discussion and validation with ASEAN Member States.

9.3.2 Potential Pilot on Climate Change Adaptation Finance

In ASEAN and other countries globally, the issue of financing the identified 'climate-smart' infrastructure arises. It is anticipated that around USD 90 trillion worth of infrastructure investments are required globally until 2030 and a significant portion must be 'climate-smart' infrastructure (WRI 2014). Despite investments through national governments, the World Bank, and development organizations, significant private sector capital will need to be mobilized. Private sector capital mobilization has also emerged as a critical issue in the UNFCCC discussions around climate adaptation finance.

The proposed pilot is to assess the feasibility and opportunities for mobilizing private sector finance options for ASEAN Member States that seek international assistance for their urban infrastructure adaptation needs. Using surveys and interviews with relevant stakeholders from key capital cities of ASEAN Member States, a framework is to be developed for identifying finance options across the range of government-led and private sector-led financing. The project would seek to identify the criteria and requirements for pursuing alternative financing options with a specific emphasis on mobilizing private sector finance, e.g. via impact investment projects, Public–Private Partnership (PPP) projects, or capital market mechanisms such as 'green bonds' or equity derivatives. ASEAN would be well suited for such a pilot project. Its member nations are faced with the broadest possible range of adaptation needs ranging from coastal infrastructure, power and water supply, biodiversity to transportation and urban building infrastructure. Such an adaptation finance framework based on 'real' climate-smart infrastructure projects would serve the needs of ASEAN Member States who require climate finance.

Several government agencies and institutions in the ASEAN Member States could be involved in the project. These include the following:

- National Construction Regulation Authorities: Primarily government agencies in charge of the development of built environments, which refers to buildings, structures, and infrastructure that provide the setting for the community's activities.
- Ministry of Finance: Finance ministries from the ASEAN Member States are responsible for administering and regulating the financial institutions and structures of the economy. Their role will be critical in the development and regulation of the proposed innovative market mechanisms, i.e.: Green Bonds, Adaptation Funds, etc.
- Scientific Research Organizations: Entities that work to advance scientific understanding of climate variability and change and its associated weather systems affecting the region, so that the knowledge and expertise can benefit decision-makers and the community.
- Urban Redevelopment Authorities: They are the central agencies responsible for land use planning. They are usually in charge of long-term strategic land use plans and medium term plans, which are reviewed every five to 10 years.

- Private Sector 'Green Finance' Departments: Various private sector banks are increasingly engaged in structuring green finance projects, such as 'green bonds' for state / municipal and corporate finance clients. For example, BNP Paribas, Citi Group, and Goldman Sachs have recently set ambitious green finance goals for their organizations.
- World Bank: The Bank uses innovative mechanisms to finance climate-friendly projects. The World Bank Green Bonds have raised funds from fixed income investors to support World Bank lending for eligible projects on climate change and adaptation in developing countries. The product was designed to respond to specific investor demand for a triple-A rated fixed income product that supports projects that address climate challenge. Since 2008, the World Bank has issued around USD 8.5 billion equivalent in Green Bonds through 100 transactions in 18 currencies.

As needed, and depending on the specifics of the infrastructure adaptation measures currently developed by the government, private sector infrastructure planning companies could also be involved in the project. These include consulting companies offering a broad range of professional services in areas of engineering, urban planning, and development and has ongoing programmes on sustainable and resilient infrastructure in the context of climate change such as ARUP and AECOM. ARUP has co-authored a report for the C40 Cities Climate Leadership Group on how to meaningfully measure climate action. AECOM has previously conducted a study on Infrastructure and Climate Change Risk Assessments for the state of Victoria in Australia taking into account infrastructure assets such as buildings, energy networks, water infrastructure, ports, roads, bridges, and communications.

Local academic sector stakeholders within ASEAN Member States for the project could also be involved. An example is the Centre for Sustainable Asian Cities at the National University of Singapore. An interdisciplinary group within NUS, the Centre serves as a platform to develop cutting-edge urban planning and design solutions as well as building technological innovations for high-density environments that will be applicable to many cities in Asia and other developing countries.

Impacts of climate change are expected to be felt on infrastructure resources in the ASEAN region (Asian Development Bank 2009). Critical infrastructure, military installations, transport infrastructure ports would be increasingly vulnerable to rising sea levels, storm surges, and flooding. Extreme weather events in adjoining countries could affect energy production, transportation, and transmission thus impacting regional energy security. Higher temperatures are also expected to result in higher summer loads and increase net electricity usage. Impact on critical infrastructure in ASEAN could also trigger migration between countries and increase conflicts over food and water. The developed adaptation finance framework would seek to ensure that both speed and scale of required adaptation investments are met and investments are financed at the lowest possible costs. From the perspective of national governments, the mobilization of private sector capital for climate change adaptation provides needed leverage to pursue other government-led investments into housing, health, social security, and defense.

Mobilizing private sector adaptation finance is a recognized global challenge. As such a pressure-tested framework and approach based on real identified adaptation needs would contribute to ASEAN's adaptation planning and offer insights for urban and non-urban adaptation finance options going forward. A pilot project to identify feasibility and opportunities for developing a climate adaptation finance framework and private sector capital mobilization would need to be supported by the lead coordinating platforms for climate change adaptation in the ASEAN Member States. Participating government agencies would then be identified as needed, depending on final scope and approach to the project. Opportunities for cash and in-kind co-financing from participating governments and private sector stakeholders have to be explored. With Singapore as a key financial hub, a broad network of expert partners in finance and academia from ASEAN Member States could be involved in the process.

Potential experts in ASEAN Member States for collaboration, knowledge sharing, and communication are the World Bank Singapore Urban Hub and C40 Cities Climate Leadership Group. The World Bank Singapore Urban Hub aims to leverage the development expertise of Singapore and the World Bank for the benefit of Asian countries. Some of its thematic areas of focus include city management, city financing, urban design, and climate change. The Urban Hub could deliver workshops and training programmes and facilitate the creation of a virtual knowledge network. The C40 Cities Climate Leadership Group is a network of the world's megacities committed to addressing climate change. C40 supports cities to collaborate effectively, share knowledge and drive meaningful, measurable, and sustainable action on climate change. Created and led by cities, C40 is focused on tackling climate change and driving urban action that reduces greenhouse gas emissions and climate risks, while increasing the health, wellbeing, and economic opportunities of urban citizens.

9.3.3 Potential Pilot on Resilient Food Supply Chains in ASEAN

Singapore being a land-scarce country imports 90% of its food from within and outside of Asia (Agri-Food and Veterinary Authority of Singapore). As a food importing country, Singapore must safeguard its food supply against possible fluctuating food prices and volatility of food supply as a consequence of climate change. Singapore is taking steps to strengthen and diversify its food supply chain to make it more resilient. In order to increase domestic production of food, Singapore launched an S$ 63 million Agriculture Productivity Fund in August 2014 (Singapore Food Agency 2006b). The Fund is expected to help local farmers purchase technology to boost local food production. About 20 farms from 17 companies have benefitted from this fund so far. The Agri-Food and Veterinary Authority is collaborating with food companies to test-bed cold chain systems like advance freezing to reduce wastage and exploring ways to recycle by-products as value-added food and animal feed.

Efforts are also on to get food importers to diversify their overseas food sources instead of focusing on regional countries to lower cost and simplify logistics.

The potential pilot proposed entails developing a holistic approach for assessing food security challenges and response options for ASEAN Member States. The pilot would develop a replicable methodology using scenario development techniques, use of regional and global climate projections, modelling of major food crop markets, and commodity prices to assess food sourcing risks and response options. The developed methodologies, tools, and processes for a fact-and-science-based assessment would be shareable and transferrable.

Climate change is regarded as one of the most pressing issues affecting ASEAN food security. Rising temperatures and sea levels are expected to impact millions of people. Rising sea level in low-lying areas would lead to inundation and salinization of agricultural land. Other impacts of climate change such as ocean acidification and warming waters are seen to already affect fish catch in the coral and pelagic areas, therefore, threatening the livelihood of fishing communities. Fish migrate as a response to warming temperatures, largely to deeper waters, which makes it more difficult for fishermen to access them as opposed to corporate trawlers. This may in turn result in the socioeconomic or political disenfranchisement of fishing communities, growing deprivation, outmigration, and food insecurity. Singapore imports fish from 46 countries of which more than 80% comes from Southeast Asia. The tropical region of Southeast Asia may lose between 40 and 60% of its fish catch potential due to climate-induced fish migrations, due to migration of fish catch away from the equator.

On land, rainfall patterns and dry periods are becoming less predictable thus affecting the rivers downstream. Increases in temperature can affect the germination and gestation process of key crops. Rice is once such key crop for the region, which has an economic, political, and nutritional significance to Southeast Asia. The nature of climate threats to rice is well understood in the context of temperature change and natural disasters. Heat stress impacts gestation periods, causes pollen sterility, and decreases overall yields. Loss of yield in low-lying areas due to changing weather patterns or salt water intrusion may necessitate the use of high lands for rice production, in spite of a lower rice yield and compromises on the production of other food commodities, such as fruits, which are traditionally more profitable.

The impacts of climate change are expected to affect ASEAN trade in agricultural products. The ASEAN Member States produce major products such as fish, fruits, and fresh vegetables, which are then traded amongst the Member States. Impacts of intra-regional trade of food products also have implications on world food prices and the region's ability to obtain food from the global food market. Asian Development Bank's International Model for Policy Analysis of Agricultural Commodities and Trade (IMPACT) projects rice prices to increase by 29–37% by 2050. Similarly, by 2050 wheat prices may increase by 81–102%, corn prices by 58–97%, and those of soybean by 14–49%.

If such price increases were to be true, it would create food security concerns for import dependent countries in ASEAN including Singapore, Indonesia, and the Philippines as net exporters like Thailand and Vietnam may opt to impose export restrictions to stabilize domestic markets as was seen in 2008. Changes in the supply of food also have implications for world prices and the region's capacity to obtain food from the global market.

This pilot project would be applicable to all ASEAN Member States. The stakeholders involved would consist of government, academia, and the private sector. Government agencies and institutions in the ASEAN Member States could also be involved in the project. These include the following:

- Agri-Food and Veterinary Authorities from ASEAN Member States: They are the government agencies responsible for ensuring food safety, resilience of food supply chains, and safety of animal and plant health, and facilitating agro trade. These are usually the go-to agencies for food related matters.
- Scientific Research Organizations: Organizations that work to advance scientific understanding of climate variability and change and its associated weather systems affecting the region, so that the knowledge and expertise can benefit decision-makers and the community. For example, the Centre for Climate Research, Singapore has been developing state-of-the-art downscaled models for Singapore and the neighbouring countries which could be used as a tool for analyzing future climate trends in a region in relation to food crops.
- Retailers like NTUC Fair Price and Cold Storage are crucial players in the food value chain in Singapore. Retailers in other ASEAN Member States can play an important role by investing resources and encouraging local production of food.
- Food processing and Storage Units: Food processing and storage units in ASEAN could play a key role by investing in local processing storage to increase the shelf life and safety standards for food during a warming and more volatile climate trend.

Local academic sector stakeholders within ASEAN Member States for the project could also be involved. An example is the S. Rajaratnam School of International Studies (RSIS), which is an institute within the Nanyang Technological University, which conducts policy-relevant research in defence, national security, international relations, strategic studies, and diplomacy. The RSIS has an active Food Security Programme under its Centre for Non-Traditional Security Studies (RSIS 2015). Other research institutes from the Consultative Group for International Agricultural Research (CGIAR) Consortium, i.e.: International Rice Research Institute, Philippines, and World Fish, Malaysia could also be involved to understand the food security challenges in their respective fields.

9.4 Concluding Remarks

The Singapore Government has taken substantial adaptation actions to protect the island state from the impacts of climate change based on solid climate science. The government views climate change adaptation as a continuous process within a whole-of-nation approach. Adaptation actions by the government have involved coastal protection, addressing flood risks, management of water resources, safeguarding biodiversity and greenery, strengthening resilience in public health and food supply, creating a cooler built environment, maintaining essential services, and keeping buildings and infrastructure safe. Two potential pilots that would benefit ASEAN as a whole were identified based on several aspects such as financial feasibility, prospective impacts on resources, and regional security implications, among others. These are on climate change adaptation finance for urban infrastructure and resilient food supply chains.

Acknowledgements The authors acknowledge the ASEAN-India Green Fund for supporting the Project on 'Enhancing Local Level Climate Change Adaptation in Southeast Asia' (Phase 1). This chapter draws partly on findings of the Project, specifically the assessment and scoping conducted in 2015 to identify the two pilots.

References

Asian Development Bank (2009) The economics of climate change in Southeast Asia: a regional review. The World Bank, Manila

BCA (2013) Inspiring change for a better tomorrow: BCA annual report 2012/2013. Building and Construction Authority, Singapore

Chen Q, Taylor D (2020) Economic development and pollution emissions in Singapore: evidence in support of the environmental Kuznets Curve hypothesis and its implications for regional sustainability. J Clean Prod 243:118637. https://doi.org/10.1016/j.jclepro.2019.118637

Government of Singapore (2010) Fourth National report to the CBD. National Parks Board, Singapore. https://www.cbd.int/doc/world/sg/sg-nr-04-en.pdf. Accessed 24 April 2016

Government of Singapore (2014) Our home, our environment, our future (sustainable Singapore blueprint), ministry of environment and water resources & ministry of national development. Singapore

Government of Singapore (2020) Singapore's update of its first nationally determined contribution (NDC) and accompanying information. Government of Singapore. https://www4.unfccc.int/sites/NDCStaging/Pages/Party.aspx?party=SGP. Accessed 28 April 2020

IPCC (2014). Climate change 2014: synthesis report. contribution of working groups I, II and III to the fifth assessment report of the intergovernmental panel on climate change. IPCC, Geneva, p 151

Li X, Wang X, Babovic V (2018) Analysis of variability and trends of precipitation extremes in Singapore during 1980–2013. Int J Climatol 38(1):125–141

MSS (2015) Singapore's second national climate change study (climate projections to 2100—report for stakeholders). Meteorological service Singapore, Singapore

NCCS (2012) Climate change and singapore: challenges, opportunities, partnerships, (National climate change strategy, 2012). National climate change secretariat. Singapore

NCCS (2015) Results from the second climate change study. National climate change secretariat. Singapore. https://www.nccs.gov.sg/news/results-second-national-climate-change-study-consistent-intergovernmental-panel-climate-changes. Accessed 24 April 2016

NEA (2018) Singapore's fourth national communication and third biennial update report. National Environment Agency, Singapore

RSIS (2015) Impact of climate change on food production: options for importing countries. https://www.rsis.edu.sg/wpcontent/uploads/2015/05/PB150529_Impact-of-Climate-Change-on-Food-Production.pdf. Accessed 26 May 2015

Sieber J, Fremgen L, Pons M (2015) Assessment of ecosystem services for urban resilience—case study in Singapore. GRF Davos Planet@Risk 3(1)

Singapore Food Agency (2006a) Food and veterinary authority of Singapore. http://www.ava.gov.sg/explorebysections/food/singapore-food-supply/the-food-we-eat. Accessed 26 July 2016

Singapore Food Agency (2006b) Agriculture productivity fund. http://www.channelnewsasia.com/news/singapore/how-cansingapore/2225088.html. Accessed 26 July 2016

World Bank (2020) World bank national accounts data and OECD national accounts data files. https://data.worldbank.org/indicator/NY.GDP.PCAP.CD?locations=SG. Accessed 19 May 2020

WRI (2014) Better growth, better climate: the new climate economy synthesis report. World Resources Institute, New York

Chapter 10
Climate Change Adaptation in Thailand

Seree Supratid and Thannob Aribarg

Abstract Thailand has a long coastline that is exposed to the impacts of climate change. The country ranks sixth in terms of the severity of projected effects of climate change due to sea level rise. The diversity of its regions is the most pronounced attribute of Thailand's physical setting, with the mountainous region located in northern Thailand. With 25 major river basins and agriculture as a gross national product, Thailand is the world's largest exporter of rice, and is often called "the rice bowl of Asia." Agriculture employs 49% of the population and contributes 10% of GDP. The focus of research in vulnerability and adaptation in Thailand is to enable the integration of climate change adaptation into national development policies and planning. The Ban Klong Sai Village of Maharaj City located in the Ayutthaya Province is proposed as a local level pilot for flood adaptation. The aim is to monitor the sustainability of integrating disaster risk reduction and climate change with respect to community livelihood, good governance, and other aspects.

Keywords Climate change adaptation · Agriculture · Vulnerability · Disaster risks · Thailand

10.1 Introduction

Natural resources make up the backbone of Thailand's economy. Agriculture is the main activity, driving the manufacturing sector and spin-offs in the service industry such as food processing and tourism. Climate change will have a great impact on the country. Thus, it is important that plans for climate change adaptation be given urgent attention.

S. Supratid (✉) · T. Aribarg
Climate Change and Disaster Center, Rangsit University, Rangsit, Thailand
e-mail: supratid@yahoo.co.th

T. Aribarg
e-mail: thannob.a@rsu.ac.th

J. J. Pereira et al. (eds.), *Climate Change Adaptation in Southeast Asia*,
Disaster Risk Reduction, https://doi.org/10.1007/978-981-16-6088-7_10

10.1.1 Climate Change in Asia

The Intergovernmental Panel on Climate Change (IPCC) reveals that climate related catastrophes during the past four decades (1950–1992) have increased about five-fold and damages to economies have increased ten-fold during the same period. There are several major conclusions on climate change in Asia from the IPCC Fifth Assessment Report (IPCC 2014) as follows:

- Continental interiors of Asia have become increasingly warmer since 1979 particularly over winter in China and over spring and autumn in north and East Asia.
- Northern and central Asia have become more wet with marked increase in rainfall between 1900 and 2005 but some areas in south Asia have become more dry.
- Water resource scarcity is *likely* to be impacted by increased variability of climate and faster thawing of glaciers.
- The extinction of many plants and animal species is expected in Asia due to the combined effects of climate change and fragmentation of habitats.
- Coastal ecosystems are expected to be significantly depleted due to the rise in sea level.
- There will be regional differences within Asia in the impacts of climate change on food production.
- Due to projected sea level rise, a million or so people along the coasts of South and Southeast Asia will likely be at risk from flooding.
- It is *likely* that climate change will impinge on sustainable development of most developing countries of Asia as it compounds the pressures on natural resources and the environment associated with rapid urbanization, industrialization, and economic development.
- Vulnerabilities of industry, infrastructure, settlements, and society to climate change are generally greater in certain high-risk locations, particularly coastal and riverine areas.

The late-twenty-first century, based on the Coupled Model Intercomparison Project Phase 5 (CMIP5) simulations under all four Representative Concentration Pathway (RCP) scenarios is shown in Fig. 10.1 (IPCC 2014). In the RCP8.5 scenario, the mean annual temperature is expected to exceed 2 °C above the baseline over most land areas in the mid-twenty-first century. The range is more than 3 °C in South and Southeast Asia to higher than 6 °C in high-latitude regions (Fig. 10.1). In this scenario, rainfall is *very likely* to increase at higher latitudes by the mid-twenty-first century and over eastern and southern areas by the late-twenty-first century. Under the RCP 2.6 scenario, changes expected are less than 2 °C and between 2–3 °C in the highest latitudes compared to the late-twentieth-century baseline. Rainfall increases are *likely* at high latitudes by the mid-twenty-first century, while at low latitudes it is *likely* that changes will remain within the levels of natural variability.

Southeast Asian countries must prepare to live in a changing climate particularly at the local or ecosystems level where impacts will be felt directly and specific

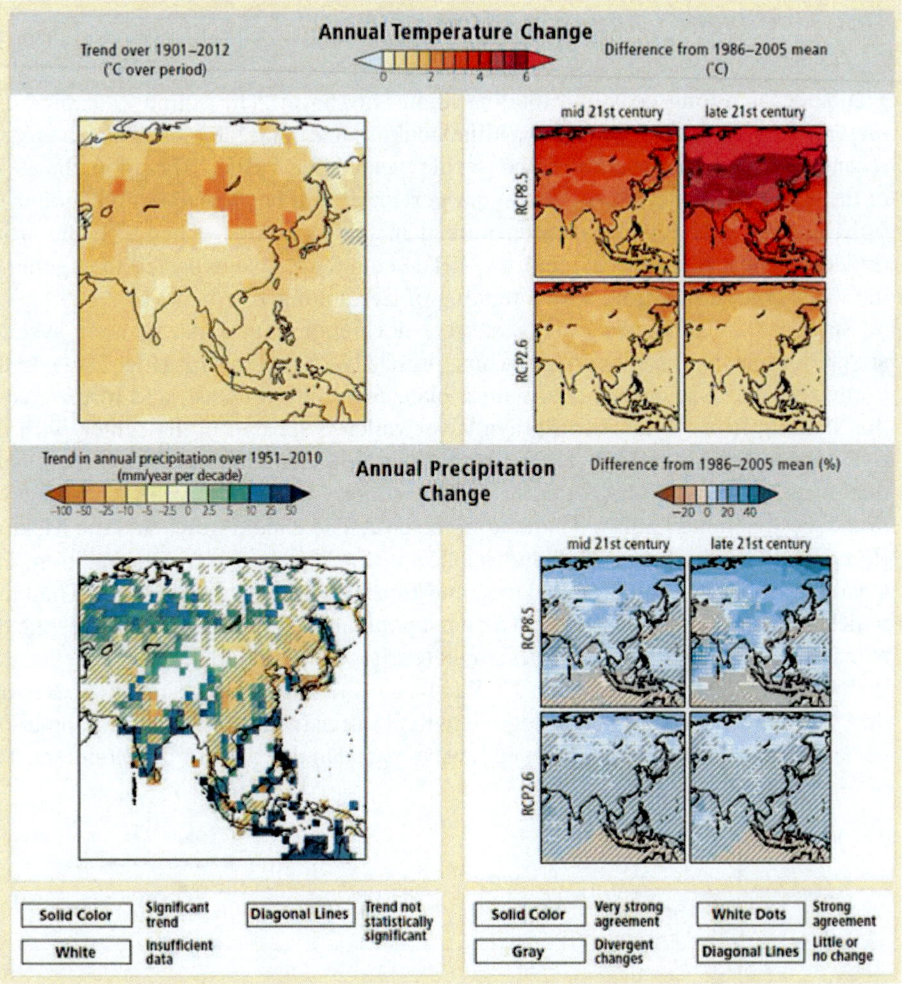

Fig. 10.1 Observed and projected changes in annual average temperature and precipitation in Asia. *Source* IPCC (2014)

actions are most needed. Failure to do so can lead to exacerbated vulnerability, resulting in deprivation, displacement, and even loss of lives in the medium to long term. Adaptation programmes will need to be planned based on inputs from pilot initiatives at the local or ecosystems level where key vulnerabilities are identified taking into account future conditions and existing capacity. Adaptation technology needs, priorities, and the potential for green growth in this context will also have to be examined.

10.1.2 Climate Change in Thailand

Thailand has a long coastline totaling about 2,615 km. The country's capital and major port is located along the coastline, making it vulnerable to climate change. A recent study ranking the cities of the world revealed that by the 2070s more than 90% of the significant asset exposed in major port cities will be located in eight countries (OECD 2007). Thailand has been ranked in the top-six to experience the worst projected impacts (Figs. 10.2 and 10.3). The country is also expected to experience the worst coastal flooding, with a ranking of five among 11 countries.

Since 2002, Thailand's water resources development and management systems are predominantly based on river basins; primarily 25 river basins (Fig. 10.4). In the north, the land is mountainous, with a plateau in the northeast, and to the east is the Mekong River. The Chao Phraya River valley is located in the center, with the river draining into the Gulf of Thailand. Southern Thailand is a narrow peninsula that opens up into the Malay Peninsula. The country is framed by the Andaman Sea in the west, and the Gulf of Thailand in the east. The Chao Phraya and the Mekong River, the lifelines of rural Thailand, support its large scale rice crop production. The Chao Phraya, Maeklong, Bangpakong, and Tapi rivers flow into the Gulf of Thailand, which extends over 320,000 km^2. The most popular tourism sites, which (despite the tsunami that occurred in 2004) still attract tourists from around the world are located in the Andaman Sea; this makes it Thailand's most precious natural resource. Besides this, there are six geographical regions which differ from the others in population, basic resources, natural features, and level of social and economic development. The

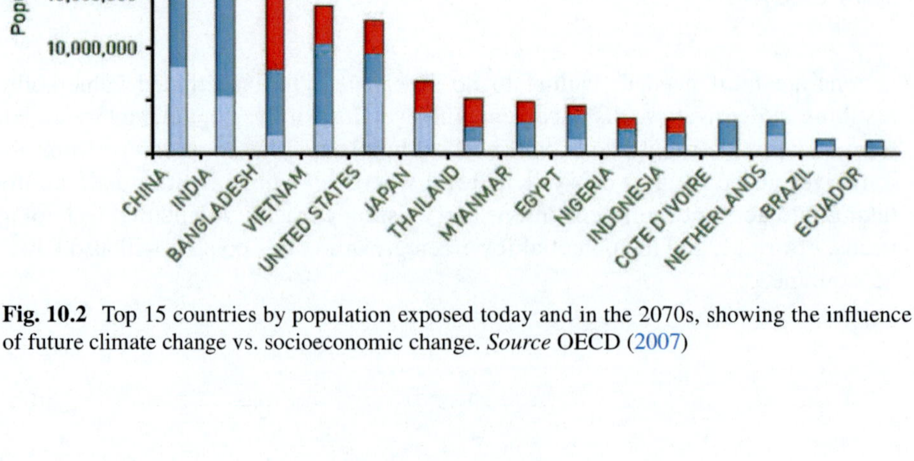

Fig. 10.2 Top 15 countries by population exposed today and in the 2070s, showing the influence of future climate change vs. socioeconomic change. *Source* OECD (2007)

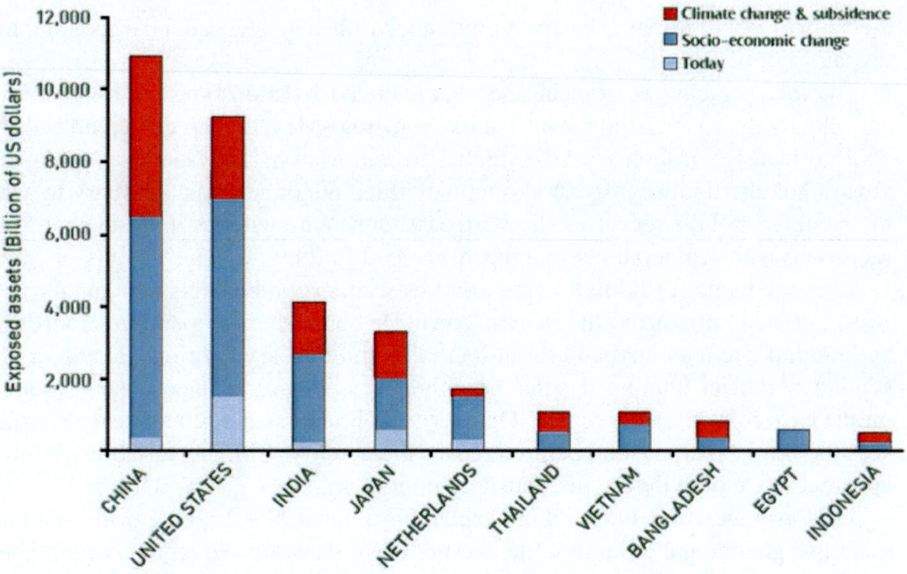

Fig. 10.3 Top 10 countries by assets exposed today and in the 2070s, showing the influence of future climate change vs. socioeconomic change. *Source* OECD (2007)

Fig. 10.4 The distribution of river basins in Thailand

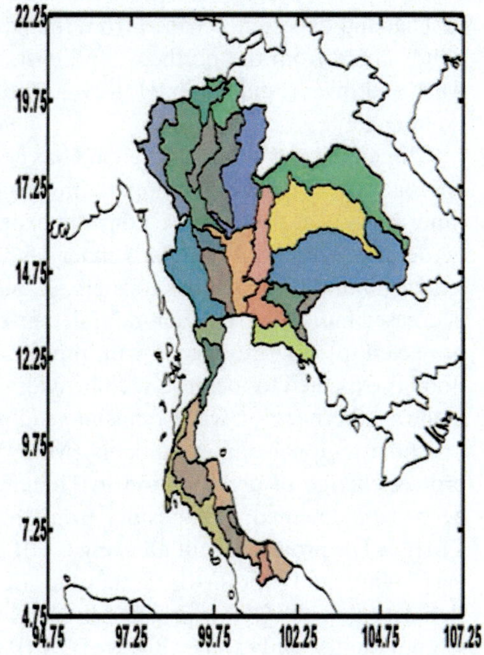

diversity of these regions is the most pronounced attribute of Thailand's geographical layout.

The local climate is tropical and characterized by monsoons. A rainy, warm, and cloudy southwest monsoon occurs from mid-May to September and a dry, cool northeast monsoon from November to mid-March. The Southern isthmus is always hot and humid. In general, there are three major seasons. February to May, the summer months, occur as the northeast monsoon switches into the southwest monsoon, with temperatures ranging from 35.0 to 39.9 °C. The months of May–October see plenty of rainfall as the southwest monsoon blankets the country with wind and rains until the northeast monsoon in October, when temperatures cool down and rainfall declines, particularly in the north and northeast regions. However, the south experiences continued rains through December, sometimes causing flooding on the eastern side of the region. October to February is the winter season, which sees the northeast monsoon begin to cover the country. Climate variability is most apparent in October during the transition period.

Total average annual rainfall in Thailand is about 1,572.2 mm. During summer, there is higher annual rainfall in the west coast of the southern region, with highest precipitation rates in September. In the winter months, the reverse is the case, with heaviest rainfall in November. It is a tropical country with high temperatures all year round and an average annual temperature of 27 °C. Maximum daily temperature of 40°C occur during the summer season. During winter months, the central parts of Thailand experience winds from the north and northeast, while in the south, the winds come from the northeast and east. The rainy season brings winds from the west, southwest, and south. However, during summer, the direction of the winds is unpredictable.

Climate downscaling in Thailand has been done by both statistical and dynamical approaches. The average temperature for the country is expected to increase over time. High heat (over 35 °C) days is expected to increase steadily in each region, while days with cool weather (under 15 °C) is projected to decrease. The average annual rainfall is expected to increase but the number of days with rainfall will decrease; indicating heavier rainfall per event. The stronger monsoon winds are projected to be more intense with repeated storm (START 2009). The risk of flash flood is expected to increase with heavier rainfall. The direction and speed of winds in the southern region with peninsula features are also expected to change (Fig. 10.5).

Three regional climate models (MRI, PRECIS, CCCMA3) have been used to project change of precipitation in Thailand (Koontanakulvong 2010). These have been supplemented with results from the multi-model ensemble of CMIP3 and CMIP5. The projected rainfall using CMIP3 (B1, A2) and CMIP5 (RCP4.5, RCP8.5) vary significantly for the different time slices compared to the period of 1961–1990. In most cases, the GCMs projects an increase in rainfall during the wet season for the whole country. Only some GCMs of CMIP3 (IPSL-CM4) and CMIP5 (IPSL-CMSA-LR) project that the rainfall will significantly decrease. Some GCMs show significant increase in rainfall for the far future (CMIP3: GISS-ER and CMIP5: CSIRO-Mk3.6). The mean projection for rainfall is generally higher for CMIP5 compared to CMIP3.

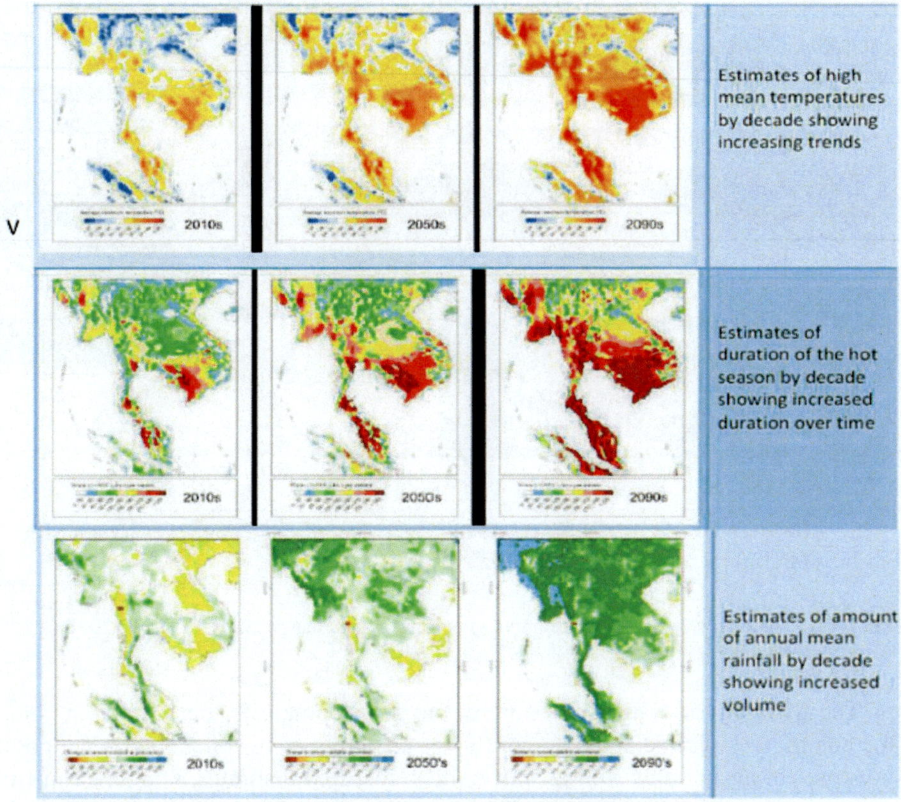

Fig. 10.5 Climate projection by dynamic downscaling (PRECIS)

10.2 National Perspective

10.2.1 National Arrangements and Institutions

Thailand ratified the United Nations Framework Convention on Climate Change (UNFCCC) in December 1994 and the Kyoto Protocol in August 2002 (MOSTE 2011). Thailand developed the Initial National Communication to the UNFCCC from 1997 to 2000, funded by the Global Environment Facility (GEF). The national focal point for climate change in the country is the Office of Natural Resource and Environmental Policy and Planning (ONEP) in the Ministry of Natural Resources and Environment (MoNRE). The Ministry of Natural Resources and Environment is the anchor for climate change in Thailand, receiving support from various departments, particularly the Office of Natural Resources and Environmental Policy and Planning. The highest policymaking on climate change is the National Committee on Climate Change, chaired by the Prime Minister and comprising three sub-committees on technical aspects, negotiation and public relations (Fig. 10.6).

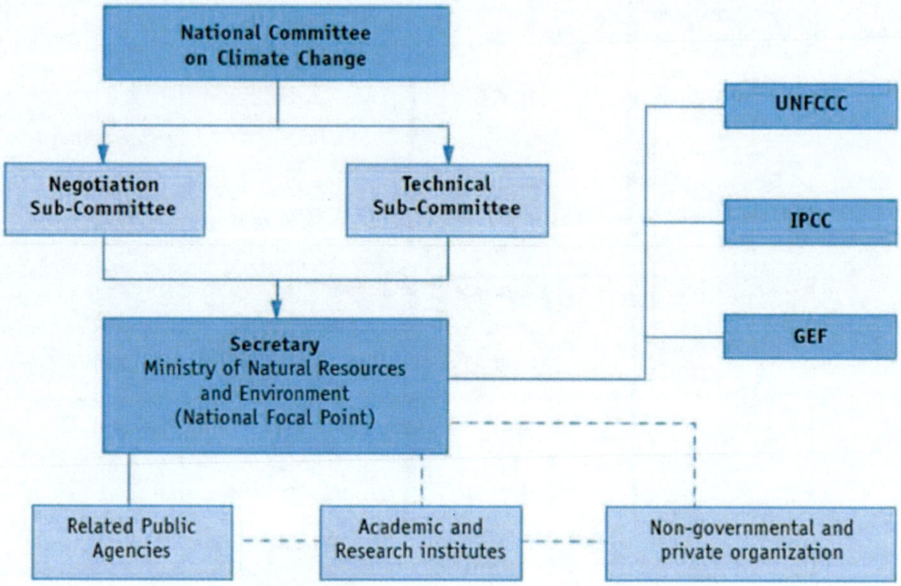

Fig. 10.6 Thailand's institutional arrangements for climate change. *Source* ONEP (2018)

The Ministry of Natural Resources and Environment is the secretariat for all the committees. The secretariat coordinates all activities in collaboration with other government agencies as well as universities, research institutes, non-governmental and private entities. In 2008, the Cabinet of Thailand approved the Strategic Plan on Climate Change (TSPCC 2008–2012). The Plan facilitates the implementation of climate change actions in the country (MONRE 2008). The Plan comprises six strategies to guide national and local agencies, where four are focused on climate change adaptation. These include building capacity to adapt and reduce vulnerability; research and development on climate impacts, adaptation and mitigation; raising awareness and public engagement; and institutional capacity building, including training of staff and establishing frameworks for coordination and integration.

Thailand's actions on climate change is also sustained by policies and plans prepared by ONEP. A 10-year Master Plan (2010–2019) on Climate Change (ONEP 2009) was put into place to meet the challenges effectively; this involved long-term development plans for all sectors: public and private, academic institutions, and common citizens (Fig. 10.7). However, there are still many challenges in the integration of disaster risk reduction (DRR) and climate change adaptation (CCA). The DRR master plan is the responsibility of the Disaster Prevention and Mitigation Department (DDPM), Ministry of Interior, while the CCA master plan is the responsibility of the ONEP, Ministry of Natural Resources and Environment (MoNRE).

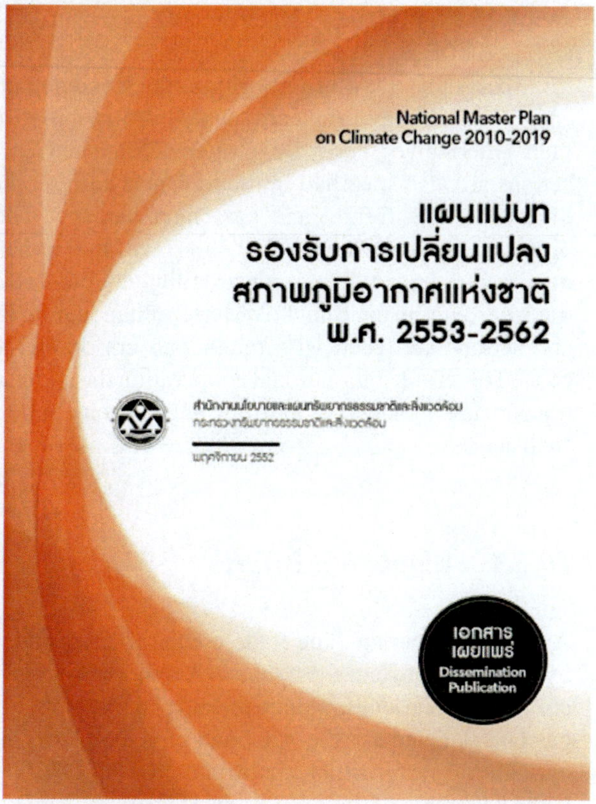

Fig. 10.7 Thailand's national master plan on climate change (2010–2019)

10.2.2 Mitigation Efforts

Thailand contributes only 0.8% of the world's carbon dioxide emissions. In 2002, the emission rate per capita of the country was 3.25 metric tons. While this is lower than the global average of 3.97 tons per capita, the levels have doubled in 2002 compared to 1991. This has served as an impetus for the government to take action. In April 2007, Bangkok hosted an IPCC Plenary Session and in the following year the city convened the UN climate change talks. The following month, the Bangkok Metropolitan Administration published the 2007 Action Plan on Global Warming Mitigation, calling for reductions in Bangkok's greenhouse gas emissions by 15% below projected 2012 levels.

Thailand submitted its Intended Nationally Determined Contribution (INDC) to the UNFCCC in 2015. Thailand's Nationally Appropriate Mitigation Action (NAMA) pledged for contribution of 7–20% GHG emission reduction by 2020 below business-as-usual level (BAU) in the energy and transport sectors. The NAMA implementation currently shows promising mitigation outcomes to ensure that the country is on track to meet the 2020 target (ONEP 2018). In May 2017, the cabinet approved the Nationally Determined Contribution (NDC) Roadmap on Mitigation (2021–2030).

Aspects that are given priority in the Roadmap on Mitigation include renewable energy and energy efficiency, which are specified under the Power Development Plan, Alternative Energy Development Plan, and Energy Efficiency Plan. Ambitious actions are also specified in the transportation sector, promoting a shift from road-to-rail for both freight and passenger haulage. The Environmentally Sustainable Transport System Plan (ONEP 2018) will incorporate these elements by expanding mass rapid transit lines, constructing double-track railways and enhancing bus transfer routes in the Bangkok Metropolitan region. However, the proposed measures and actions are faced with major barriers of very high investment and operating costs. Therefore, international cooperation through the UNFCCC such as, technology transferring, financial supporting, and capacity building are imperative to implement such actions.

10.2.3 Adaptation Efforts

Agriculture, tourism, and trade are the main economic sectors that will be affected by climate change in the country. A majority of Thailand's population of 65 million are in rural, agricultural areas. Being the world's largest exporter of rice, Thailand has been labeled as "the rice bowl of Asia." Forty-nine per cent of Thais are in the agricultural sector which contributes 10% of GDP. Thailand's 3,200 km of coastline is the backbone of the tourism and fisheries sectors; it contributes to the livelihoods of 10% of the population and 6% of GDP. Fifteen percent of the country's population live in the capital city, Bangkok, which is the economic, political and social center; not only for Thailand but also for the greater Mekong region, making it a global city.

The country's economy, culture and the environment stand to suffer tremendously as a result of climate change. Climate change and its effects include higher surface temperatures, floods, droughts, severe storms, and sea level rise. Thailand's rice crops can be wiped out by just one degree of warming, and rice production is the backbone of its economy. It is predicted that in 20 years, Bangkok is likely to be submerged due to rising sea levels; a rise by just a few centimeters will severely affect coastal tourism. In an effort to save the country's economy from the effects of climate change, Thailand is working to put into place comprehensive adaptation plans. These include a slow shift to organic agriculture, a tsunami warning system along the Andaman Sea, the construction of a flood prevention wall around Bangkok, and an action plan to reduce greenhouse gas emissions from vehicles and energy use.

In Thailand, CO_2 emission scenarios formed the basis of climate change impact and adaptation studies. (Thailand Second National Communication). This model-based approach generally covers many decades with high uncertainties. Conventional scenario planning is found to be weak in supporting climate change adaptation planning because of its static exploratory nature, lack of local level multi-sectoral inputs and relatively short time frames (e.g., 20 years), which contribute to uncertainties (Butler et al. 2020). These scientific and technical limitations have largely been ignored previously, partly due to rising occurrences of climate change-related natural disasters and the urgent and practical need to identify areas exposed to such risks. Climate variability and extreme events, such as severe droughts or floods covering a few years (5–10 years) are now being increasingly studied.

Since 2000, the Thailand Research Fund has supported much of the research work on the effects of climate change (Table 10.1). Climate change impact, vulnerability, and adaptation have been prioritized thanks to Thailand's National Strategy on Climate Change, 2008–2012. Considerable efforts have been made over the past decade to expand technical knowledge and integrate the results into the process of sustainable national development. It should also be noted that very few bilateral and multi-lateral cooperative variability and adaptation (V&A) projects, through the UNFCCC, have been promoted over the same period.

The ultimate goal of vulnerability and adaptation efforts is to integrate climate change adaptation needs into national development policies and planning. The potential impacts of climate change affect different sectors in the country. Furthermore, the adaptive capacities of communities in different geographic settings vary. The impact of climate change and vulnerability to climate change are locality-specific, making adaptation responses context-specific. The agriculture, water resources, and health sectors are more vulnerable to climate change impacts, while coastal ecological resources are vulnerable to sea level rise (ONEP 2018).

10.2.4 Vulnerable Sectors and Regions

Concepts and terminology introduced in the IPCC Third Assessment Report of 2001 have served to guide adaptation in Thailand. The method of unequal weights (Fig. 10.8) was used to construct the vulnerability index for Thailand (Prasertsak 2011). A set of indicators were selected for each of the three components of vulnerability. The data was arranged in the form of matrix and normalized using functional relationship. Vulnerability indices were computed using regional and provincial level data. Indicators were selected based on availability of data, previous research and their underlying reasons of vulnerability. Data in 2006 was used as the baseline data to construct the vulnerability index. Changes in climatic variables in 2006 were compared to the average of 30 years. It was found that, at the regional scale, Northeast is the most vulnerable to climate change. It is characterized as a very highly vulnerable region. Other regions like the North, East, and South were less vulnerable regions.

Agriculture: Agriculture is an important sector in the Thailand's economic and social development; it produces rice, sugarcane, cassava, rubber, and maize, and provides raw materials for many attendant industries (ONEP 2018). It was the source of food security and employment during the 2000 and 2007 economic crises. However, the ratio of agriculture to GDP is dropping. Notwithstanding, a majority of the population

Table 10.1 Selected projects on vulnerability and adaptation. *Source* ONEP (2018)

Project	Institution	Project period
Under Thailand research fund		
Preliminary study of climate change impact on food security in Thailand	Chiangmai University	2003–2004
Climate change and climate variability	King Mongkut's Institute of Technology Ladkrabang	2007
Climate model for Thailand and neighboring countries	Southeast Asia START	2007–2008
Downscaling global climate scenarios for Thailand	Ramkamhaeng University	2007–2009
Development of climate scenarios for Thailand by MM5 model	Chiangmai University	2007–2009
Study and development of climate model for Thailand using Reg CM3	King Mongkut's University of Technology Thonburi	2007–2009
Testing and development of weather research and forecasting (WRF) in climate forecasting for Thailand	Kina Mongkut's University of Technology Thonburi	2007–2009
Climate variation due to oceanic disturbance in Thailand	Burapha University	2007–2008
Study of aerosol effects on climate change in Thailand	Silpakorn University	2007–2009
Assessment of climate extreme events in Thailand: vulnerability and risk analysis in critical areas	Chulalongkorn University	2007–2008
Effects of climate change on monthly rainfall and surface flow in Thailand and the effects on water management in Eastern Thailand	Chulalongkorn University	2007–2009
Impact of global warming on rice, sugar cane, cassava, and maize production in Thailand	Khon Kaen University	2008–2009
Economic assessment of climate change impact on rice production in nd	Kasetsart University	2008
Impact of climate change on thermal comfort and energy consumption in residential buildings in Thailand	Kasetsart University	2008
Approaches to conduct research and to develop human resources for global warming	Thailand Environment Institute	2008
From other sources		
Enabling activities II* (UNDP-GEF)	Office of Natural Resource and Environmental Policy and planing	2002
Southeast Asia regional vulnerability to changing water resources and extreme hydrological events* (AIAAC/UNEP)	Southeast Asia START	2004

(continued)

Table 10.1 (continued)

Project	Institution	Project period
Impact of climate change on inflow of bhumibhol and sirikit reservoirs, northern Thailand (national research council)		

Note *refers to projects in English, otherwise in the Thai language

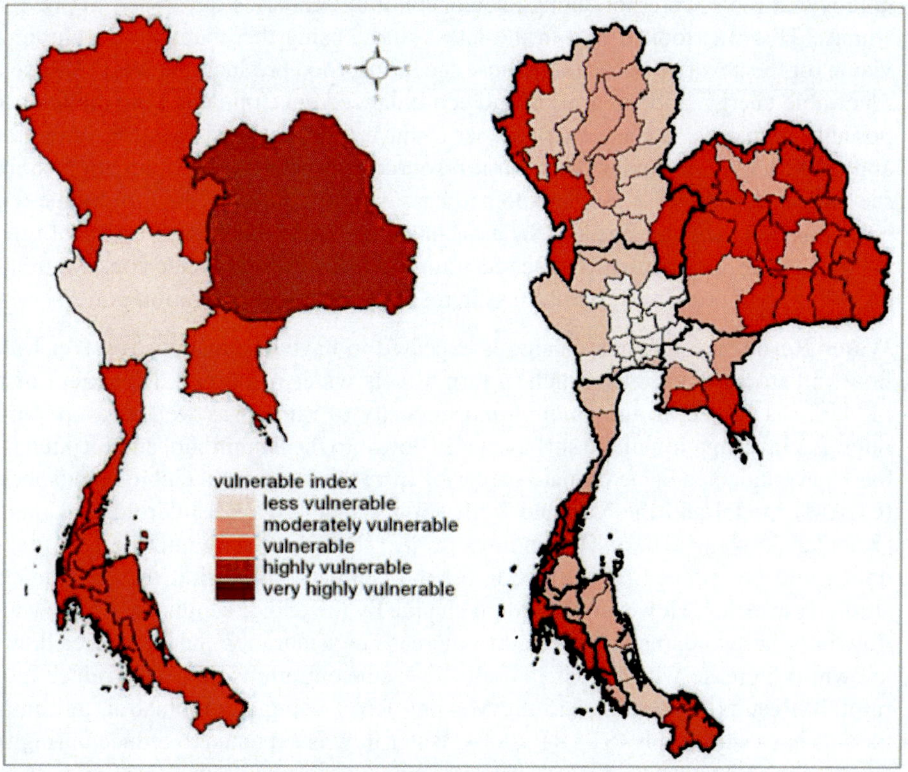

Fig. 10.8 Vulnerability to climate change in Thailand. *Source* Prasertsak (2011)

is still engaged in this sector. Floods and droughts are dominant hazards for the agriculture sectors in Thailand (ONEP 2018); however, the effects of climate change on this sector could vary. For example, in the Songkhram River Basin, combined top-down and bottom-up approaches revealed the future maximum and minimum temperatures are expected to increase by 2.8 and 3.2 °C, respectively, under the RCP8.5 scenario, with a shift in rainfall patterns (Boonwichai et al. 2019). With the help of such information, strategies were identified to produce more rice in the Basin. However, different models were used in selected areas to study the potential impact on rice and maize yields, and these showed substantial variations in production levels.

Such variations are due to the high level of uncertainties associated with the models. This problem of unreliable models has yet to be resolved.

The effects of climate change have been projected in a study which shows that the impact on agricultural crops could be moderate, even minimal. (ONEP 1999). However, climate change could cause yields of crops to fluctuate substantially. During the rainy season, northern regions could experience more problems in production; the effects during the hot season are projected to be the same across the country. However, there is level of uncertainty as only one model was used. Another issue to note is that there were not enough socioeconomic scenarios developed in previous studies. This was looked into in the latest study, using the country's development vision for the next two decades. In these scenarios, food production (World Kitchen), renewable energy, and integration and eco-balanced development were the focus of possible scenarios. Each scenario is set to have different land uses for agriculture and other activities. Under the natural resource constraints, rice production should see an increase by 23%. Sugarcane and cassava areas are expected to double with emphasis on renewable energy (Sahassachai et al. 2002). The conservation of forest is expected to increase carbon sequestration while soil and water conservation is projected to improve in agriculture systems, as part of the integration policy.

Water Resources: Climate change is expected to have implications on river water flow and storage capacity, which in turn affects water resources. The extent of its influence is subject to the amount and intensity of rainfall as well as evaporation rates. An investigation of the surface water flow into the Bhumibhol and Sirikit dams has been conducted using climate scenarios from the Conformal Cubic Atmospheric (CCAM) model and the Variable Infiltration Capacity (VIC) hydrological model (Koontanakulvong 2010). The findings project that water flow into the Bhumibhol dam could be reduced by the middle of the century. The Sirikit dam will not be similarly affected. However, this could change by the end of century when the water flow into the two dams is projected to increase considerably. Surface water flow is known to increase with higher rainfall; high concentrations of CO_2 produce more rainfall. Results from studies in the Mekong River using a combination of climate models have shown this (START 2006). Water flow is expected to reduce during the dry season. However, increased water is expected to results in flash floods in the eastern region, according to some models (START 2006).

Health: The Initial National Communication documented the first study on the impacts of climate change on health (ONEP 2018). A technical relationship between temperature and the growth rate of mosquitoes showed that global warming will result in the spread of malaria within the first half of the century. However, there has been no follow-up study over the past 10 years. A more recent study on risks of malaria and dengue and climate factors in various provinces failed to establish a clear relationship (Koontanakulvong 2010); more studies are needed. The first National Strategic Plan on Environmental Health (2008–2011) is a result of cooperation between key environmental as well as public health agencies. Approved by the cabinet on 8 December 2009, the Plan comprises six strategies, including climate change measures and approaches (ONEP 2018). The strategy emphasizes enhancement of an effective warning system to control and prevent airborne diseases caused by climate change from turning into an epidemic. The Charter of Environment and Health, signed by 14 Ministers of Health from Southeast and East Asia in 2007, provided guidance for this initiative. The Charter is expected to facilitate regional collaboration, specifically for enhancing adaptive capacity in the health sector as well research and development.

Forest and Wildlife: Since the last study in the Initial National Communication, no other studies on the impacts of climate change on forests have been carried out (ONEP 2018). The type and structure of the forest and its ecosystem, as well as the biological diversity of fauna and flora which is vital to the development of agricultural genetic resources, are the main areas of concern; these are addressed in the National Strategic Plan on Climate Change (2008–2012).

Marine and Coastal Resources: With a coastline of more than 2,600 km, the country has an ecosystem that is vital to the socioeconomic growth of the southern and eastern regions. Unfortunately, there have not been many studies done on the potential effects of global warming on the sea level. Dynamic Interactive Vulnerability Assessment (DIVA) was used in a small study along the coast of Krabi and it shows that in 25–30 years the sea level is likely to rise by 11–22 cm; it will also affect about 10–35 m of the coastline which will be inundated (START and WWF 2008). However, the sea level rise in the Gulf of Thailand over the past 56 years has shown a decline by 35 cm per century; this reverse trend could be due to geological change (Vongvisessomjai 2006). Aspects of vulnerability to sea level rise, physical changes and biological conditions have to be incorporated into socio-economic scenarios to model future impacts.

Vulnerable Regions: Climate change is challenging as it affects people's livelihoods severely. Dealing with these changes is not easy when scientific information and expertise on this subject nationally is often quite limited; currently it is mostly derived from global level knowledge, rather than from the specific country. Most of the Asian countries depend on predictions from experts in meteorology or geography. The data available on the potential trends is often generic. The mean annual temperature with a longer summer and more daytime temperatures rising above 33 °C have been forecasted. There will be fewer cold days, however, rainfall will be more intense. Water shortage and increase in drought days and flood frequency in some river basins are expected. Changes in rice productivity are also expected. There may be more wet season crops in some areas and less in others. Wetlands will be badly affected with a decrease in water. Severe coastal erosions and changes in the accretion patterns will ravage the coastal zones; already 23 coastal provinces in Thailand have lost 599 km^2 or 21% of 2,667 km. The Bangpakong and Maeklong coastal ecosystems are the most at risk (more than 25 m per year). Depending on the setting, the probability of landslide events is expected to increase sharply in high and medium altitude regions due to increase in extreme rainfall (Komori et al. 2018).

10.3 Adaptation Priorities

10.3.1 Approach

The Sendai Framework for Disaster Risk Reduction 2015–2030 was adopted by UN Member States on 18 March 2015 at the Third UN World Conference on Disaster Risk Reduction in Sendai City, Miyagi Prefecture, Japan. It is a 15-year voluntary, non-binding agreement that recognizes that the State has the primary role to reduce disaster risk but that responsibility should be shared with other stakeholders, including local government. The Sendai Framework was also adopted by the Thai Government on March 31, 2015. This has resulted in a revised National Master Plan for Disaster Prevention and Mitigation.

In Thailand, adaptation activities tend to emphasize the prevention of natural disasters; not the management of climate change. International organizations such as the MRC, UNDP, and the ADB have provided technical and financial support in the area of climate change. A study is currently underway on "Climate Change Impact Adaptation and Mitigation in Asian Coastal Mega Cities," which includes Bangkok and Ho Chi Minh City. The study is being funded by by the ADB, World Bank and other parties. The investigation on climate change issues in Bangkok and Ho Chi Minh City is also being carried out.

10.3.2 Potential Local Level Pilot

About 35% of the total land area of the country makes up the Chao Phraya river basin, which makes it the largest basin in Thailand; it has an area of 159,000 km^2. Two main rivers cut across the delta: the Tha Chin River on the west and the Chao Phraya river (the main stream) on the east. Four large tributaries feed the basin: the Ping, Wang, Yom, and Nan which originate from the mountainous terrain in the northern part of the country. These four tributaries flow southward to merge in Nakhon Sawan to form the Chao Phraya River which flows on southward through a large alluvial plain into the sea at the Gulf of Thailand (Fig. 10.9). The basin area is flat with an average elevation of 1–2 m from the mean sea level (MSL). Some parts are below the sea level as a result of land subsidence. The basin is dissected by canals. Bangkok is located 33 km uplands from the Gulf of Thailand along the Chao Phraya River. The area is exposed to annual flooding from rivers from the north and inundation due to the high tide from the sea as a result of proximity to the sea and flat geography.

The worst flood ever recorded in the country was in 2011 (Fig. 10.10). Compared with Hurricane Katrina, this was a more expensive disaster; estimated damage and losses came up to USD 45.7 billion (World Bank 2012). Although efforts have been made to mitigate the flood damage in the Chao Phraya River Basin through several structural measures (construction of dams, reservoirs, dikes and pumping stations), flooding still has much more impact as a result of deforestation, farmland expansion

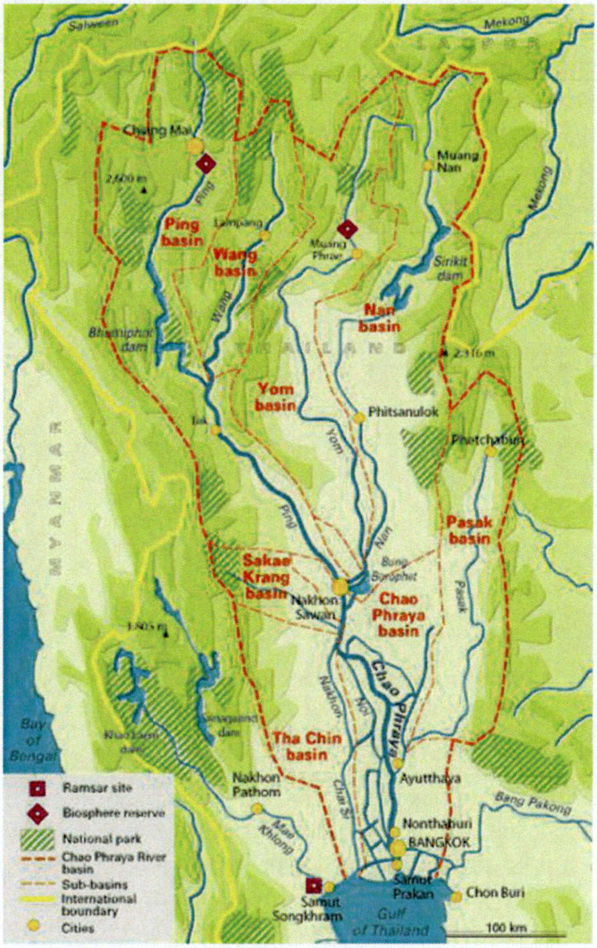

Fig. 10.9 The Chao Phraya river drains from the north of Chiang Mai to the Gulf of Thailand in the south

and urban development. Ayutthaya was one of the most heavily impacted areas, where all five industrial estates were inundated in 2011 (Supharatid 2015). The soil barrier along the Chao Phraya River had been breached at many points upstream from Chainat to Singburi Provinces. This caused floodplain flow to both sides of the river and downstream flow directed to several industrial areas. In addition, the limited carrying capacity of the Chao Phraya River, especially at Ayutthaya, also caused more impact on the downstream industrial areas. The flood damage potential is increasing due to rapid urbanization and land development in downstream areas; particularly in Ayutthaya and its municipalities along the Chao Phraya River.

A small village, "Ban Klong Sai" in Maharaj City (red star in Ayutthaya Province) is proposed as a local level pilot for flood adaptation. In this pilot, the concept of sustainable development that comprises the three pillars of economic (eco-friendly), social (people-centred and participation), and environment (green community) will be integrated with DRR and CCA approaches.

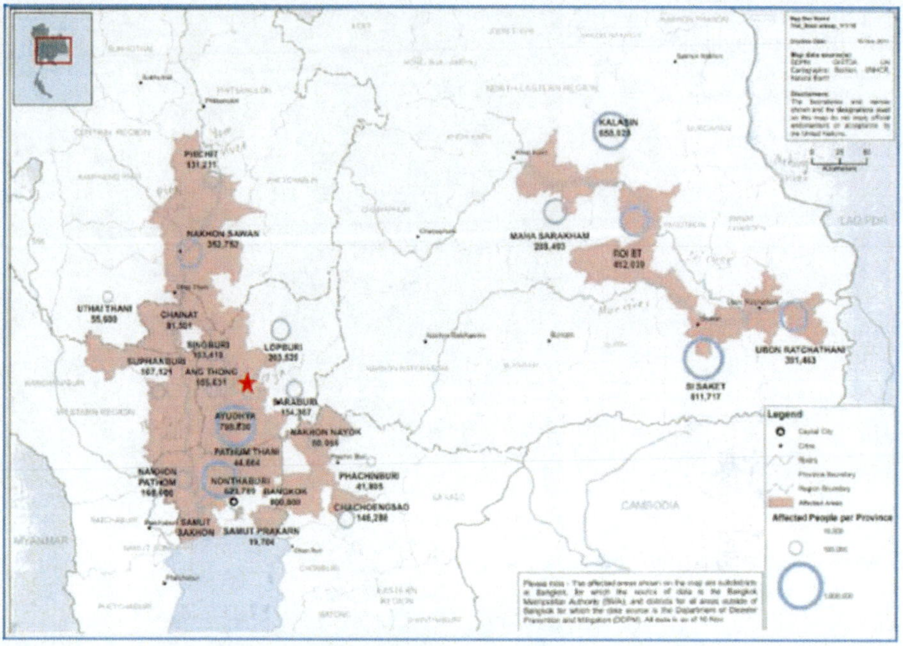

Fig. 10.10 The extent of areas affected by floods in 2011

The focus of the pilot will be on a project that has commenced; to build houses for 22 families badly affected by the major floods in the Ayutthaya Province, with a design concept to enable sustainable living with floods. This is one of the several public–private partnership projects initiated in the Province since the 2011 floods. In this project, all the houses are 3 m above the ground to avoid flood damage, and the houses are connected by 3 m-high boardwalks; thereby, reducing flood vulnerability. There is new sanitary system that can be used in times of floods, reinforced with concrete structure and flood-resistant fiber cement. The design concept of this "cool house" also reduces electricity consumption by natural shading and air ventilation. Solar energy for public lighting and energy-saving light bulbs in all houses also help to reduce greenhouse gas emissions. However, the sustainability of the project in integrating DRR and CCA approaches has to be monitored, particularly with respect to community livelihood and good governance, among others.

10.4 Concluding Remarks

Adaptation activities in the country have focused on the prevention of natural disasters rather than climate change management. However, the increase in climate change-related disasters calls for critical and practical actions to identify areas exposed to such risks. Studies to find ways to deal with changes and extreme events such as severe droughts or floods covering about 5–10 years, are now being increasingly utilized. This is primarily because model-based approaches cover many decades with high uncertainties. In addition, the lack of local level multi-sectoral inputs and the need for decision-making input in relatively short timeframes contributes to the overall weakness of conventional scenario-planning for climate change adaptation.

The agriculture sector is very important for Thailand. The findings of preliminary studies on projecting the impacts of climate change on rice and maize yields using different models are inconsistent. The uncertainties associated with these models are too high for the findings to be used by decision-makers for making policies. While the climate models have been refined continuously, problems of high uncertainties persist. This limits the use of climate change models for projecting impacts in the agriculture sector.

An emphasis on local level pilots that embeds climate change impacts modelling and iterative scenario planning within decision-making cycles, in conjunction with stakeholder engagement, could reveal new approaches to address this challenge. The Ban Klong Sai Village of Maharaj City located in the Ayutthaya Province is proposed as a local level pilot for flood adaptation. In this pilot, the sustainability of integrating disaster risk reduction and climate change will be monitored and evaluated, particularly with respect to community livelihood and good governance, among others. The learnings from this process will facilitate reflective improvements for flood adaptation in the long term.

Acknowledgements The authors acknowledge the ASEAN-India Green Fund for supporting the Project on "Enhancing Local Level Climate Change Adaptation in Southeast Asia" (Phase 1). This chapter draws partly on findings of the Project, specifically the local level needs assessment and scoping exercise conducted in 2015.

References

Boonwichai S, Shrestha S, Weesakul MS, Datta SA (2019) Evaluation of climate change impacts and adaptation strategies on rainfed rice production in Songkhram River Basin, Thailand. Sci Total Environ 652:189–201

Butler JRA, Bergseng AM, Bohensky E, Pedde S, Aitkenhead M, Hamden R (2020) Adapting scenarios for climate adaptation: practitioners' perspectives on a popular planning method. Environ Sci Policy 104(2020):13–19. https://doi.org/10.1016/j.envsci.2019.10.014

IPCC (2014) Climate change 2013: the physical science basis. In: Stocker TF, Qin D, Plattner GK, Tignor M, Allen SK, Boschung J, Nauels A, Xia Y, Bex V, Midgley PM (eds) Contribution of working group I to the fifth assessment report of the intergovernmental panel on climate change. Cambridge University Press, Cambridge and New York, p 1535

Komori D, Rangsiwanichpong P, Naotatsu Inoue N, Ono K, Watanabe S, Kazama S (2018) Distributed probability of slope failure in Thailand under climate change. Clim Risk Manag 20:126–137

Koontanakulvong S (2010) Thai climate change and its impacts towards water sector. Chulalongkorn University Thailand, Bangkok (in Thai)

MONRE (2008) Thailand strategic plan on climate change (TSPCC 2008–2012). Bangkok, Thailand

MOSTE (2011) Thailand second national communication under the United Nations framework convention on climate change. Office of natural resources and environmental policy and planning, Bangkok, Thailand

OECD (2007) Ranking of the world's cities most exposed to coastal flooding today and in the future. Newark, USA. https://www.oecd.org/dataoecd/16/10/39721444.pdf. Accessed 28 July 2016

ONEP (1999) Impacts of climate change on different sectors. Ministry of natural resources and environment. Bangkok, Thailand

ONEP (2009) National master plan on climate change (2010–2019). Ministry of natural resources and environment. Bangkok, Thailand

ONEP (2018) Thailand's third national communication to the United Nations framework convention on climate change. Ministry of natural resources and environment. Bangkok, Thailand. https://unfccc.int/documents/181765. Accessed 21 March 2019

Prasertsak (2011) Country report: Thailand, workshop on climate change and its impact on agriculture. Republic of Korea Seoul, 13–16 Dec 2011

Sahassachai K, Vinai S, Sukij R (2002) Impact of climate change on maize, sugarcane and cassava in KhonKaen Province. Northeastern Thailand

START (2006) Southeast Asia regional vulnerability to changing water resource and extreme hydrological events due to climate change (AIACC AS07). Southeast Asia START regional center technical report no.15. Bangkok

START (2009) Future climate protection for Thailand and mainland Southeast Asia using PRECIS climate models. Southeast Asia START regional center, technical report no. 18. Bangkok

START and WWF (2008) Climate change impacts in Krabi Province, Thailand: study of environmental, social and economic challenge. Southeast Asia START regional center and world wildlife fund, Bangkok

Supharatid S (2015) Assessment of CMIP3-CMIP5 climate models precipitation projection and implication of flood vulnerability of Bangkok. Am J Clim Chang 4:140–162

Vongvisessomjai S (2006) Will sea-level really fall in the Gulf of Thailand? Songklanakarin J Sci Technol 28(2):227–248

World Bank (2012) Thai food 2011: overview rapid assessment for resilient recovery and reconstruction planning. The World Bank, Bangkok, pp 1–24

Chapter 11
Climate Change Adaptation in Vietnam

Trong Dinh Tran and Thang Van Nguyen

Abstract Vietnam is exposed to frequent and intense hydro-meteorological hazards such as storms, floods, and droughts. The proposed local level adaptation pilot is the Quang Ngai Province, which has experienced very high incidences of typhoons, floods, riverbank and shoreline erosion, mountain landslides, and flash floods. The coastal areas are also exposed to sea level rise, drought, and saline intrusion. Loss and damage due to sea level rise is linked to fishery productivity, degradation of mangrove forests, saline intrusion in brackish and freshwater aquaculture zones, and inundation of transportation infrastructure. Adaptation has been given serious attention in the country. However, lack of capacity and resources are major barriers. There is a need for capacity building, transfer of technology, and financial support to advance adaptation in Vietnam.

Keywords Climate change adaptation · Mitigation · Disaster risks · Mekong River · Vietnam

11.1 Introduction

Vietnam is situated in Southeast Asia, stretching from latitudes 8°27 to 23°23N and longitudes 102° 08 to 109°30E. The country borders China to the north, Laos to the northwest, and Cambodia to the southwest, with a coastline of 3,260 km and land area of 330,992 km^2. Vietnam has a coastal system of about 3,000 big and small islands with a total area of more than 1,600 km^2. Three of four Vietnam territories are covered by mountains and hills with the elevation mostly from 100 to 1000 m. The plains are concentrated downstream of two big rivers: The Red and Mekong rivers. The river network in Vietnam is rather dense with 2,360 rivers. The average density is 0.6 km/km^2. The topography of Vietnam comprises seven regions: (i) Northern mountain region; (ii) Red river delta; (iii) Northern central coast; (iv) Southern central

T. D. Tran (✉) · T. Van Nguyen
Vietnam Institute of Meteorology, Hydrology and Climate Change, Hanoi, Vietnam
e-mail: trantrong209@gmail.com

J. J. Pereira et al. (eds.), *Climate Change Adaptation in Southeast Asia*,
Disaster Risk Reduction, https://doi.org/10.1007/978-981-16-6088-7_11

coast; (v) Central highland; (vi) North-eastern of the south; and (vii) Mekong river delta (General Statistical Office 1996).

11.1.1 Current Climate

Vietnam has a tropical monsoon climate with plentiful heat and humidity. However, it stretches across 15 degrees of latitude, making its topography diverse. Therefore, there are marked differences in climate across Vietnam. Annual mean temperatures have been recorded to range from 18 to 29 °C. The mean temperature for the coldest month is about 13–20 °C in the mountainous north and 20–28 °C in the south. Summer temperatures range from 25–30 °C. In terms of precipitation, the annual rainfall figures range from 600 to 5,000 mm with about 80–90% of rainfall recorded during wet season. The annual number of rainy days ranges from 60 to 200 days with floods and inundation occurring in some regions. Drought is often recorded during the dry season because of uneven distribution of rainfall. Vietnam, being located as it is in the northwest Pacific Ocean, experiences an average of 4–5 typhoons or tropical cyclones annually.

Over the past five decades, Vietnam has seen an increase of approximately 0.5 °C in average temperature with a sea level rise of about 20 cm. Extreme climate events like storms, floods, and droughts have increased both in frequency and intensity (MONRE 2008). Analysis of data indicates changes of climate parameters and sea level with some distinct features.

Temperature: The last 50 years saw a rise of about 0.5–0.7 °C in the annual average temperature in the country. Winter temperatures went up faster than summer temperatures. Similarly, temperatures in the northern climate zones also experienced a faster increase than those of the southern climate zones (Fig. 11.1). The annual average temperature for the last four decades (1961–2000) was higher than that of the three previous decades (1931–1960). For the period of 1991–2000, Ha Noi, Da Nang, and Ho Chi Minh City saw annual average temperatures higher by 0.8, 0.4, and 0.6 °C, respectively, than the average for 1931–1940. In 2007, the annual average temperatures at these three locations were all higher by 0.8–1.3 °C than the average for 1931–1940; and by 0.4–0.5 °C for the period of 1991–2000 (MONRE 2008).

Rainfall: The changes in annual average rainfall across all regions over the last 90 years (1911–2000) have not been distinct nor consistent. Records show ascending and descending periods with decreasing annual rainfall over the north and increasing rainfall in the south (Fig. 11.1). On average, Vietnam has seen a drop of about 2% in rainfall over the past 50 years (1958–2007) (MONRE 2008).

Cold Front: The country has experienced remarkably fewer cold fronts in the last two decades. However, there have also been frequent anomalous extreme events. For example, in the northern region during January and February 2008, a cold surge lasted for 38 continuous days causing considerable damage (MONRE 2008).

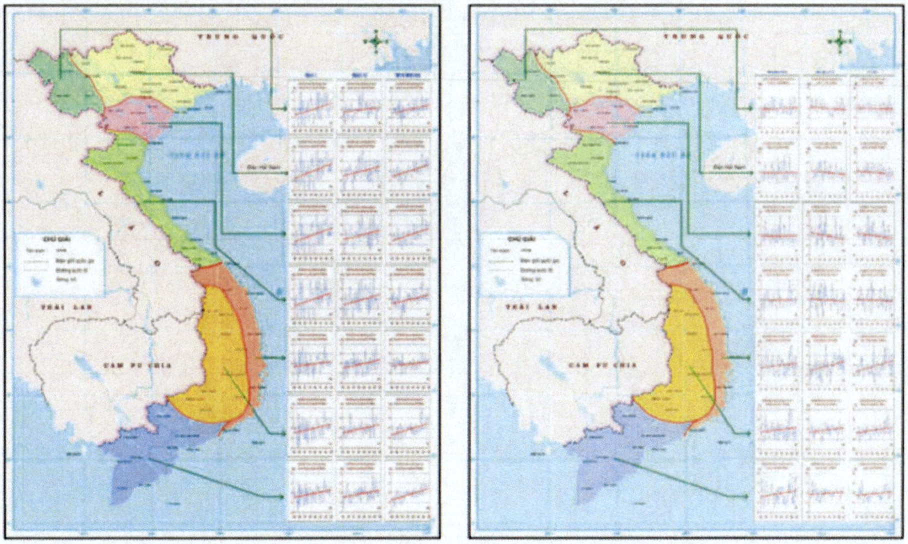

Fig. 11.1 Trend of temperature (left) and rainfall (right) over Vietnam for the last 50 years

Typhoon: Over the recent years, there have been more occurrences of higher intensity typhoons; they tend to move southward and last longer. Also, they have shown more abnormal movements (Fig. 11.2).

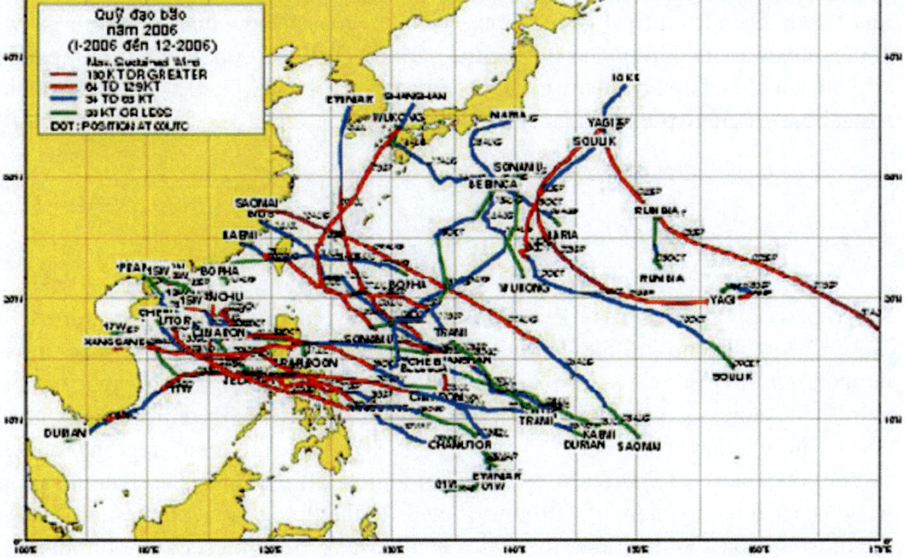

Fig. 11.2 Typhoon tracks in the North-Western Pacific Ocean and East Sea. *Source* MONRE (2003)

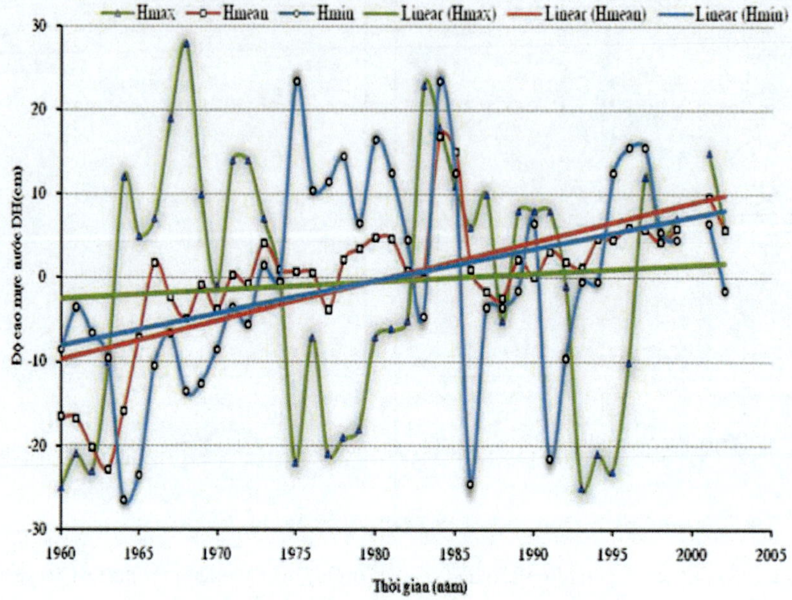

Fig. 11.3 Changes in sea level at Hon Dau oceanographically station

Drizzle: There has been a drop in the average number of days with drizzle in Ha Noi over the period of 1981–1990. In the past decade, drizzle days reduced to only half (15 days/year) of the long-term average (Nguyen and Nguyen 2004).

Sea Level: Data from tidal gauge along Vietnam coasts show that sea level rise was at the rate of about 3 mm/year during the period of 1993–2008 which is comparable with the global tendency. In the past 50 years, sea level (Fig. 11.3) at Hon Dau station rose about 20 cm (MONRE 2008).

11.1.2 Projected Climate

Large parts of Vietnam may see an increase in average annual temperatures over 2–3 °C by the end of the twenty-first century (MONRE 2012a). The lowest temperature may increase from 2.2–3 °C average while the highest average temperature may increase from 2–3.2 °C. The number of days with temperatures above 35 °C (maximum) is expected to rise by 15–30 days over many regions. Annual rainfall is expected to increase by 2–7% in almost all regions. The overall rainfall trend is expected to change, with lower levels in the dry season and higher levels in the rainy season. Compared to the 1980–1999 levels, the highest daily rainfall levels are expected to increase in the northern part of the country. Abnormal rainy days with levels twice the current record are expected in some areas. The rainfall levels are expected to decrease in the southern part of the country (MONRE 2012a).

The end of 2100 may see significant changes in climate extremes with projected increases in average minimum (2.2–3.9 °C) and maximum temperature (2.0–3.2 °C). The Northeast region of Vietnam and the south of Central highland areas expect a relatively higher increase in temperatures than that of other regions. The number of days with maximum temperature over 35 °C is projected to increase from 15 to 30 days over most parts of Vietnam and the maximum values of daily rainfall are likely to increase in North to North-Central regions of Vietnam, decrease in the South-Central, Central Highlands, and southern regions of Vietnam.

Average sea level rise may be in the range of 57–73 cm Vietnam, the highest sea level rise may be in the region from Ca Mau to Kien Giang Provinces with 62–82 cm, and lowest in the region from Mong Cai to Hon Dau with 49–64 cm.

11.1.3 Socioeconomic Setting

Vietnam has various economic sectors. These include the state economy, collective economy, private economy (individual owner and private capital), and foreign investment economy. One of the measures that the Government of Vietnam took to implement and mainstream the state economic sector is to establish collective State economic groups and state corporation (SOEs). Since the early 1990s, Vietnam has consistently promoted equitation of the SOEs. Although the State policy development gives priority to the state economic sector and collective economic groups, the growth rate of these two components is lower than the private sector and foreign capital investment. The land valuation and mapping system is also currently being reviewed from multiple dimensions for the purpose of helping the government to draw up policies and regulate land rights (Chinh et al. 2020).

In the agriculture sector, the focus is on improvements in the quality of cultivated products and meeting the demands for local as well as export requirements. Cultivated areas and production of some industrial crops show increase, whereas rice and corn production meet the demand for food consumption and food security in the country, with a significant part for exports. Large-scale industrial farming is the direction that animal husbandry is moving toward. However, diseases, long periods of cold weather, and other disasters have been serious hurdles in this sector over the recent years. A decreasing trend has been noted in the area of aquaculture water surface too, but the total value of fish and fishery production increased.

In the forestry sector, the total forested area in 2010 was about 13,388,100 ha with 39.5% coverage. Natural forests comprised 10.3048 million hectares and while plantations covered 3.0833 million hectares. From 2008 to 2012, forest area and forest coverage rate increased. The total value of forest production at constant prices since 2010 defined by sector activity increased, specifically 17,202.3 billion in 2008, to 18,714.7 billion in 2008, and 21,136.0 billion in 2012.

The proportion of industry in the national economy remains low but the value of industrial production at constant prices since 2009 by economic sector (such as State-owned, Private, and Private sector with foreign investment) has increased (General Department of Statistics 2012). Fourthly, transportation industry has developed rapidly in all aspects but still not enough to meet the increasing demands of society. Road transport accounted for the highest percentage of transportation, followed by the waterways. The total number of passengers and volume of freight has increased over the years. For example, in 2008, the total number of passengers was 1793.5 million; in 2010, it was 2315.2 million; in 2012, it rose to 2775.9 million. Similarly, in 2008 the total volume of cargo rose from 653.23 million tons to 800.88 million tons in 2010, and 959.30 million tons in 2012 (General Department of Statistics 2013).

In the area of energy, the country has taken steps to diversify energy sources, promote energy saving and efficiency; it has also issued policies related to energy efficiency and renewable energy (MONRE 2012b). Eighty percent of the energy consumption by the people living in the mountainous areas is derived from biomass sources such as wood, agricultural residues, rice husk, and biogas partly. By promoting and developing the use of biogas, logging activities in natural forests can be minimized. Some facilities which produce electricity from renewable energy sources are being connected to the national grid to supplement the power supply.

Being the highest body of authority in the country, the National Assembly can draw up, adopt and amend the constitution, and formulate new laws or amend existing ones. The National Assembly has decision-making powers on socioeconomic development plans and budgets and provides oversight on the performance of executive bodies.

11.2 National Perspective

11.2.1 National Arrangements and Institutions

Over the past 10 years, national policies and measures to mitigate greenhouse gasses (GHGs), and adapt to climate change have been successfully initiated and funded by domestic resources. Furthermore, Vietnam supports achieving a legal agreement in conjunction with all Parties that have ratified the United Nations Framework Convention on Climate Change (UNFCCC). Like many other countries, it is working toward keeping the global average atmospheric temperature increase at below 2 °C.

A National Steering Committee was set up to implement the UNFCCC (signed in 1992 and ratified in 1994) and Kyoto Protocol (signed in 1998 and ratified in 2002); subsequently, the Initial National Communication (2003), the Second National Communication (2010), and the Initial Biennial Update Report (2014), reflecting the latest climate change response efforts and GHG inventories were duly submitted to the UNFCCC Secretariat.

In 2008, the Government issued the National Target Programme to Respond to Climate Change (NTP-RCC). Climate change was mainstreamed into the National Socioeconomic Development Strategy (2011–2020) and Socioeconomic Development Plan (2011–2015) enabling policies on disaster risk reduction, coastal zone management, and energy supply and use, to be put into place. Economic sectors and provinces have since developed Action Plans to respond to climate change.

The National Climate Change Strategy was formulated in 2011. The objectives for 2011–2015 and 2016–2050, and the priority projects for the period of 2011–2015 were outlined. Response to climate change is linked to sustainable development and transition toward a low-carbon economy, to take advantage of opportunities to increase competition and strengthen the national position, and carry out adaptation and mitigation efforts in parallel.

The National Green Growth Strategy was launched in 2012 (GoV 2012). Mitigation goals and actions, and regulations to align with international carbon markets were laid out. In 2013, the Law on Natural Disaster Prevention and Control was passed, and climate change appeared as a whole chapter in the 2014 Law on Environment. However, both human and financial resources are needed to ensure the success of all these policies and activities.

The Intended Nationally Determined Contribution (INDC) of Vietnam includes mitigation and adaptation components. In addition to reporting actions being taken, the adaptation component highlights adaptation gaps. These gaps are related to institutional and policy arrangements, financing, human resource capacity and technology, and prioritized adaptation measures for the 2021–2030 period. The national budget is only likely to accommodate about one third of the financial needs for this period; international support and private sector investment have to be sought for the rest.

Proactive adaptation actions will lessen vulnerability and inequality within and among countries. Such measures will help Vietnam build resilience, and do its own part in reducing GHG emissions. Climate change adaptation measures must be well-thought out and planned to address urgent and immediate impacts as well as potential long-term impacts. The goal is to achieve sustainable development and gradually ease the country into a low-carbon economy. A systematic, joint, interdisciplinary, inter-regional approach is needed; one that also promotes gender equality, eradicates hunger, and uplifts the poor. The various strategic measures and action plan advocate proactive adaptation to climate change at the national and local levels with linkages to disaster risk prevention and reduction (GoV 2011; 2013).

11.2.2 Near and Long-Term Climate Change Adaptation

Climate change adaptation has been linked to disaster risk management as reflected by the numerous activities implemented at the national level. The costs associated with the impacts of climate hazards, specifically sea level rise and saltwater intrusion, are expected to increase but resources are limited at the national level. It is estimated that adaptation costs will exceed 3–5% of the GDP by 2030 (MONRE 2012a). Funding for adaptation is required from multiple sources including the public and private sectors as well as international agencies.

The adaptation programs and projects at the national level have been being carried out mainly from projects supported and the state budget (GoV 2011; MONRE 2012b; 2019). The programs and projects include:

- National Target Programme to Respond to Climate Change (NTP) with the goal of (i) assessing the impacts of climate change (including climate variability, sea level rise, and extreme climate events) for the areas and sectors and local authorities, (ii) Identification of measures to respond to climate change, and (iii) Develop and implement an action plan to respond to climate change with a focus on adaptation.
- Supported Program to Respond to Climate Change (SP-RCC) in order to support the development and implementation of policy and institutional framework, the investment projects on climate change.
- National Strategy on Climate Change identified 10 key tasks, whereas the Action Plan implementing the national strategy on climate change, including 65 schemes and projects implemented in 2012–2020, of which 25 projects are adapting to climate change.
- National Strategy for prevention and disaster risk reduction by 2020 with specific actions, harmonious coordination between non-structural measures and structural measures are considered to be the basis for steering and agreement to act consistently and vigorously at all levels and the community.
- The program "Raising public awareness and disaster risk management community based", planned for 12 years for 6,000 communes and villages frequently affected by natural disasters focuses on education and raising awareness of the community, with a total budget of VND 988.7 billion, of which the central budget is VND 546.9 billion (up 55%), community contributions make up VND 45.322 billion (5%), and loans (ODA) from countries and international organizations are VND 395.48 billion (up 40%).

Long-term climate change adaptation covers the period 2021–2030, which was essentially determined on the outcomes of near-term programs as well as analysis of future projections. Continuing the long-term task of the national strategy on climate change adaptation measures in this period are proactive measures to cope, monitor, provide early warning, mitigation, food security, and water resources, sustainable economic and social development MONRE (2019). The measures are grouped into three major programs covering disaster risks, social security, and sea level rise (Table 11.1).

Table 11.1 Proposed programs and measures for long-term climate change adaptation in Vietnam. *Source* MONRE (2019)

Program for Disasters and Climate Monitoring
• Modernize the hydro-meteorological observatory and forecasting system to ensure the timely forecasting and early warning of weather events; • Develop the assessment and monitoring system on climate change and sea level rise; • Produce socioeconomic development plans based on climate change scenarios, with a focus on key sectors and regions; • Implement disaster prevention plans and measures, protect peoples' lives, and ensure national defense and security; • Consolidate and develop prioritized and urgent disaster prevention projects; strengthen the capacity of search and rescue forces; • Develop infrastructure and make plans for residential areas; relocate and resettle households and communities from areas affected frequently by storm surges, floods, riverbank and shoreline erosion, or areas at risk of flash floods and landslides; • Allocate and mobilize resources for community-based actions, raising awareness and building capacities for climate change adaptation and disaster risk management
Program for Ensuring Social Security
• Review, adjust and develop livelihoods and production processes that are appropriate under climate change conditions and are linked to poverty reduction and social justice; • Develop mechanisms, policies, and strengthen the insurance system, and share climate and disaster risks; • Improve regulations and technical standards for infrastructure, public facilities and housing, that are appropriate under climate change conditions; • Implement ecosystem-based adaptation through the development of ecosystem services and biodiversity conservation, with a focus on the preservation of genetic resources, species at risk of extinction, and important ecosystems; • Implement community-based adaptation, including using indigenous knowledge, prioritizing the most vulnerable communities; • Implement integrated water resource management in river basin systems; ensure reservoir safety; strengthen international cooperation in addressing transboundary water issues; ensure water security; • Ensure food security through protecting, sustainably maintaining and managing agricultural land; restructuring of crops and livestock; create new climate change resilient varieties; complete the disease control and prevention system; • Implement sustainable forest management; improve the quality of poor natural forests; implement afforestation and reforestation measures, focusing on large timber plantations; prevent forest deforestation and degradation; • Protect, restore, plant and improve the quality of coastal forests, including mangroves, especially in coastal estuaries and the Mekong and Red River deltas
Programs for Sea Level Rise and Urban Inundation
• Implement integrated coastal zone management; • Use sea level rise scenarios in urban and land use planning for infrastructure, industrial parks, coastal and island resettlement areas; • Implement anti-inundation measures for large coastal cities; construct climate change resilient urban infrastructure; strengthen and build new large urban drainage infrastructure; • Consolidate, upgrade, and complete crucial sea and river dykes; • Control saline water intrusion in the most severely affected areas

11.2.3 Vulnerable Sectors and Regions

By the year 2100, an increase of 2–3 °C is projected in the annual average temperature while precipitation may rise during the wet season and drop in the dry season (ADB 2013). Furthermore, sea level is expected to rise of between 78 and 100 cm. Extreme climate events will increase in both frequency and intensity. Climate change will result in more intense hazards, especially storms, floods, and droughts.

Without doubt, climate change and its attendant disasters are expected to have severe effects on Vietnam, with the most vulnerable sectors being agriculture, natural ecosystems, biodiversity, water resources, public health, and infrastructure (MONRE 2019). The most vulnerable regions are the Mekong Delta, the Red River Delta, and the Central Coast. The Mekong Delta is one of the deltas in the world most susceptible to sea level rise. The poor, ethnic minorities, the elderly, women, children, and people with disabilities are the most vulnerable.

There will be dire consequences if adaptation measures are not put into effect. For example, sea level rise of 100 cm will put of many parts of the country at risk of inundation (MONRE 2012b; 2019). Areas at risk include over 10% of the Red River Delta and Quang Ninh Province that will directly affect 9% of the total population. More than 2.5% of the area of the central coastal provinces is also at risk potentially impacting 9% of the population. Furthermore, up to 39% of the Mekong Delta could be submerged, affecting 35% of the population. This also leads to a potential loss of 40.5% of the total rice production in this region. Increases in temperature are expected to affect the mountain regions in the north affecting agriculture (crop, livestock, and fisheries), natural ecosystems, biodiversity, energy security, and public health (MONRE 2012b; 2019).

Tropical cyclones and flooding is projected in the coastal zones including the Mekong Delta. The risk of landslides is expected to increase in the mountain regions of the north. Anomalies arising from climate change as more complex and cause unexpected consequences (MPI 2011; 2012). Climatic extremes are different in terms of intensity and spatial distribution across Vietnam, so the role, impact, and danger level of each hazard, and the exposure to extreme risks vary and depend on the nature of affected areas (Table 11.2).

Despite putting into place much needed actions to minimize the effects of climate change, the country will still suffer major losses and damages; the damage to residential areas, economic zones, and ecosystems will be beyond its resilience and capacity to handle. The solution lies in considering sharing and managing these risks of loss at both the national and international levels. Technical measures can be taken to prevent future losses. However, such measures, for example, to mitigate the effects of floods, storm surges, saline water intrusion, and drought, have to be implemented with some urgency, within this century. This country does not have the capacity to do so currently as the national budget will only cover a third of the financial needs to implement adaptation measures in the country (GoV 2016).

Table 11.2 Regional vulnerabilities caused by disaster and climate change in Vietnam. Exposure levels are defined as 4: extreme danger; 3: danger; 2: moderate danger; 1: less danger; 0: safe. *Source* World Bank (2010)

Region	North West	North East	Red river delta	North central	South central	Central highland	East of the South	Mekong river delta
Exposure								
Storms	1	3	4	4	4	2	2	3
Floods	1	1	4	4	4	2	2	4
Saline Intrusion	0	0	1	2	2	0	1	4
Sea Level Rise	0	0	2	2	2	0	3	4
Drought	2	2	1	4	4	4	2	2
Average	**1.2**	**1.5**	**2.2**	**3.2**	**3.2**	**1.7**	**1.8**	**3.0**
Vulnerability								
Property	4	3	2	4	2	4	1	2
Economic diversity	4	4	2	4	3	4	2	2
Education	4	3	1	2	2	2	1	3
Health and Hygiene	4	1	2	1	1	1	1	3
Ethnic Minorities	4	3	0	1	1	4	1	2
Women and children	4	3	1	2	3	3	1	2
Migrants	0	0	2	2	1	4	4	1
Urban households	0	0	2	1	1	0	4	3
Average	**3.0**	**2.1**	**1.5**	**2.1**	**1.8**	**2.8**	**1.9**	**2.3**
Total	**4.2**	**3.6**	**3.7**	**5.3**	**5.0**	**4.5**	**3.7**	**5.3**

11.3 Adaptation Priorities

11.3.1 Key Issues

Vietnam has drawn up policies, and put into place measures, and carried out activities to adapt to the effects of climate change. However, the country just does not have sufficient capacity and resources. To achieve its objectives, capacity building, technology transfer, and financial support need to be mobilized.

Capacity Building: There are not enough experts and technical staff at the local level with the necessary skills in this field (MONRE 2019). Better communication and awareness-raising programmes are needed. Forecasting disasters and early warning capacities, as well as scientific research and adaptation technology are still limited. Adaptation approaches at the community level need to be identified and replicated. The capacity to select and decide on prioritizing resources for the adaptation implementation is also limited.

Technology: There is a shortage of advanced technologies for hydrological and meteorological monitoring and forecasting, early warning of natural disasters and hazards, and climate change adaptation (MONRE 2019). For example, the country requires technology for real-time forecasting, early warning, and sharing information system on real-time hydro-meteorological monitoring. In addition, tools to assess climate change impacts, vulnerability, exposure, and climate change adaptation measures are also required. Technology for the sustainable use of water resources, prevention of water pollution, and undisrupted urban water supply is also critical.

Finance: While there are policies, plans and programmes, climate change adaptation efforts were designed to collect funding for implementation; state resources can only meet 30% of the adaptation needs (GoV 2016). Financing is needed mainly to maintain existing infrastructure, construct much needed projects to prevent natural disasters, and for socioeconomic development. For example, an assessment of climate change impacts on the drainage in Quy Nhon City has revealed the urgent need to improve urban drainage design in the cities (Vo et al. 2020). Vietnam encourages and creates favorable conditions for private sector investment in climate change adaptation activities.

11.3.2 Potential Local Level Pilot

The Quang Ngai Province is proposed as the local level pilot. The Province faces annual storms and flooding resulting in frequent sedimentation, aquatic breaks and inundation in the lowlands, coastal erosion, and landslides in mountainous areas, all of which affect agriculture, forestry and fisheries, and livelihoods; making it difficult to manage and use the land (Quang Ngai Department of Agriculture and Rural Development 2011a; b). The Quang Ngai Province is located in the south-central coast; thus, there is a high possibility that it will also be affected by sea level rise. Coastal plain estuaries such as the east of Binh Son district (Sa Can estuary), the east of Son Tinh district, Tu Nghia, Mo Duc (Cua Dai, Lo estuary), and southeast of Duc Pho district (My A, Sa Huynh estuary) are highly vulnerable to sea level rise.

Geographic Setting: The Province is situated in the coastal area of the Central Region, with geographical coordinates "from 14°32′04 to 15°25′00" North latitude and "from 108°14′25 to 109°09′00" East longitude (Fig. 11.4). The coastal line of over 140 km covers the East with many large estuaries such as Sa Ky, Cua Dai, My A, and Sa Huynh. The topographic features of Quang Ngai are mountain terrain encroached into seaside with transitional terrain from coastal plains in the east to the mountainous terrain in the west (Quang Ngai Department of Planning and Investment

Fig. 11.4 Administrative map of the Quang Ngai Province

2010). With elevations of 1500–1800 m in the western mountains and from 5–30 m in the plains, topography forms a high and low lying landscape.

Current Climate and Drainage: Quang Ngai has a typical tropical climate with high temperatures, and plenty of sunlight, precipitation, humidity, and wind regime. With high annual total temperatures and the annual rainfall amounts in Quang Ngai, it is favorable for crop diversification (Pham and Phan 1993). The river network in Quang Ngai is evenly distributed across the Province and they originate from the mountains in the west (Tran et al. 2011). The slopes of short rivers are relatively large (>2%). The downstream of the rivers are influenced by tides and salt water intrusion and there is strong siltation and erosion in estuaries along the river. Flow diversion is intense for all the rivers downstream and in the dry season, the water in the rivers is almost exhausted. There are four rivers, i.e., Tra Bong, Tra Khuc, Ve and Tra Cau, which have a major impact on the economic development and livelihoods of people in the province. Overall, river water in Quang Ngai is unevenly distributed between the months of the year, and year to year. The number and magnitude of floods are determined by the upper rainfall centers in the districts of Tay Son, Son Ha, Ba To, Tay Tra and Tra Bong. The rivers of Quang Ngai provide water supply for agriculture, industry, and the people's livelihood. There is no significant transportation in the waterway.

Climate Projection: Climate change is already impacting various socioeconomic aspects of the Quang Ngai Province, especially in the agriculture, forestry, water resources, and industry sectors. According to some recent calculations, the agricultural land at risk of being flooded by impacts of climate change and sea level rise (scenario A2 period 2080–2099) are 31,977 ha, equivalent to 6.21% of the provincial area and 23.5% of total agricultural land. In particular, the areas of paddy land at risk of being flooded are 28,890 ha which is equal to 39.76% of the total paddy land. In the forestry sector, the changing climatic factors such as temperature, rainfall, CO_2 concentration, or monsoon seasons can have positive or negative impacts on the growth and diversity of plant species as well as the quantities and species of animals. In the industry sector, the rise in temperatures can directly impact this sector. The high temperatures affect the working environment of workers. Machinery and equipment are also more easily damaged in hot weather conditions and thus, more money will have to be spent on maintenance or new purchases. Consequently, the cost of products can increase, which can then decrease the competitiveness of enterprises and the revenue. The change in rainfall could directly affect industrial production activities. Heavy rain could hinder the process of transportation and distribution of goods, damage property, and increase the product prices. Increase in rainfall during the rainy season could also cause flooding in industrial facilities over prolonged time that may lead to production disruptions, damage to equipment, machinery, and serious environmental pollution by emitted toxic chemicals.

Awareness Level: The people's awareness about climate change in Quang Ngai has been improved by the support from provincial government and other sectors, but the ability to respond and adapt to climate change is still limited. In order to enhance that ability of Quang Ngai in any specific stage for sustainable development, it requires suitable orientations, policies, and projects aimed at improving the awareness and the response to climate change of the community, social organizations, and government departments. The policies and action plans to respond to climate change should be integrated into each strategy of the socioeconomic development of provinces, districts, communes, and other sectors.

Institutions and Co-financing Potential: The Provincial People Committee of Quang Ngai has established the Steering Committee on Responding to Climate Change (Fig. 11.5). The organizational structure for implementing climate change adaptation is led by a Chairman, who is Vice President of Provincial People Committee, with support from Standing Vice Chairs from the Department of Natural Resources and Environment, Department of Planning and Investment, and Department of Finance. The Head of Water Resource, Hydro-Meteorology Division serves as Secretary while Commissioners are drawn from the Provincial Department and Sectors, including Industry and Trade, Science and Technology, Agriculture and Rural Development, Construction, Transportation, Information and Communications, Culture-Sports and Tourism, Agriculture and Rural Development, Education and Training, Labor, Invalids and Social Affairs and the Center for

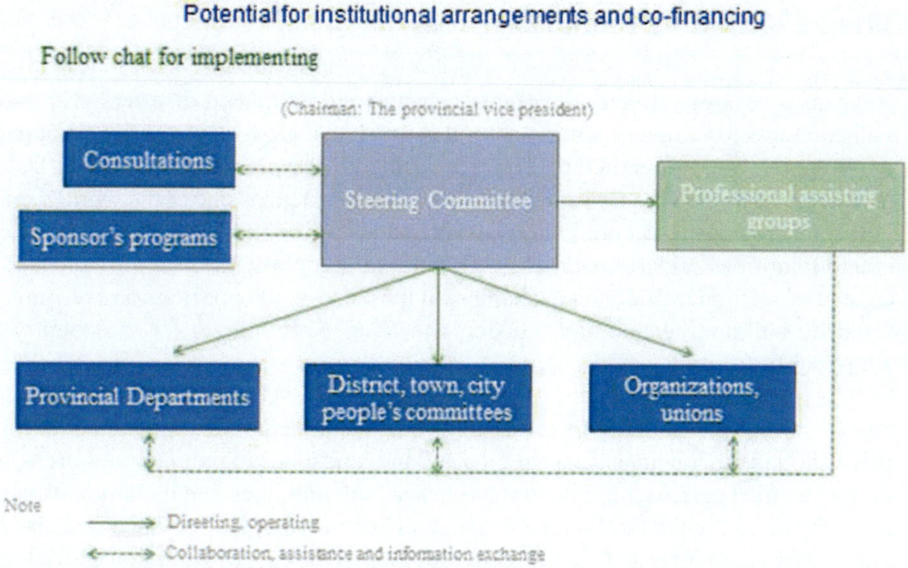

Fig. 11.5 The steering committee on responding to climate change in Quang Ngai Province

Hydro-meteorology, among others. The co-financing approach is to effectively use internal resources, and international capital sources based on the following general principles (i) enlist international capital (non-refundable aid, loans) with the state budget, the state credit budget and People's capitals; (ii) use appropriate capital structure with budget support and stimulate other sources of capital; (iii) encourage and create favorable environment to attract investments from enterprises and private sources in any form; and (iii) use the tax resources / environmental fees from businesses and from the use of natural resources.

Partnership Potential: At the national level, partnership developments would require a system of policies that integrate climate change into sectoral programs and specific tasks; strengthening and improving the system of organization, identifying the responsibilities of relevant agencies and funds, and mechanisms to manage the tasks of the action program to mitigate and adapt to climate change. The Quang Ngai Province has experience in efficiently partnering with international donors and leveraging on external finance and technology transfer through collaborative activities. With the aim of attracting international investment and collaboration, the Provincial People Committee have taken the following actions: (i) developed mechanisms for cooperation between the Province and international donors to implement the programs and projects of the Action Plan; (ii) expanded relationships and cooperation in responding to climate change with international donors and organizations; and (iii) formulated legal documents for the province on mechanisms and policies to encourage investment in the Clean Development Mechanism projects, and environmentally friendly technology transfer to facilitate foreign partners to invest in the Province.

11.4 Concluding Remarks

Vietnam is severely affected by climate change and its related disasters. The most vulnerable sectors are agriculture, natural ecosystems, biodiversity, water resources, public health, and infrastructure. The most vulnerable regions are the Mekong Delta, Red River Delta, and the Central Coast. The Mekong Delta is one of the world's most vulnerable regions; susceptible to sea level rise where the most affected are the poor, ethnic minorities, elderly, women, children, and people with disabilities. Climate change adaptation in Vietnam is challenged by shortage of capacities and resources. Capacity building, technology transfer, and finance are critical for the country to address future impacts.

The Quang Ngai Province is proposed as the local level pilot. The Province faces annual storms and flooding resulting in frequent sedimentation, aquatic breaks and inundation in the lowlands, coastal erosion, and landslides in mountainous areas, all of which affect agriculture, forestry and fisheries, livelihoods and making it difficult to manage and use the land. There is a high possibility that the Province will also be affected by sea level rise. Coastal plain estuaries in the east of Binh Son district and southeast of Duc Pho district are highly vulnerable to sea level rise.

Acknowledgements The authors acknowledge the ASEAN-India Green Fund for supporting the Project on "Enhancing Local Level Climate Change Adaptation in Southeast Asia" (Phase 1). This chapter draws partly on findings of the Project, specifically the local level needs assessment and scoping exercise conducted in 2015.

References

ADB (2013) Viet Nam environment and climate change assessment. Asian Development Bank, Mandaluyong City, p 1–68

Chinh NT, Song NP, Manh LV, Thuy NT, Vu TN, Hanh NT, Linh LP (2020) Factors that affect land values and the development of land value maps for strengthening policy making in Vietnam: the case study of non-agricultural land in Quang Ninh Province Vietnam. EQA-Int J Environ Qual 36:23–35

General Department of Statistics (2012) Statistical yearbook 2011. Ha Noi, Vietnam

General Department of Statistics (2013) Statistical yearbook 2012. Ha Noi, Vietnam

General Statistical Office (1996) Statistical Yearbook 1995. Statistical Publishing House, Ha Noi, Vietnam

GoV (2011) National strategy on climate change. Government of Vietnam, Ha Noi, Vietnam

GoV (2012) National strategy on green growth. Government of Vietnam, Ha Noi, Vietnam

GoV (2013) Resolution no. 24 / NQ-TW—Pro-actively responding to climate change, enhancing natural resource management and environmental protection. Government of Vietnam, Ha Noi, Vietnam

GoV (2016) Intended nationally determined contribution of Vietnam submitted to the United Nations framework convention on climate change. Ha Noi, Vietnam

MONRE (2003) Vietnam initial national communication to the United Nations framework convention on climate change. Ha Noi, Vietnam

MONRE (2008) National target program to respond to climate change (NTP) 2008–2015. Ha Noi, Vietnam
MONRE (2012a) Climate change scenarios, sea level rise for Vietnam. Vietnam publishing house of natural recourses, environment and cartography, Ha Noi, Vietnam
MONRE (2012b) National target program to respond to climate change in the period of 2012–2015. Ha Noi, Vietnam
MONRE (2019) Vietnam third national communication to the United Nations framework convention on climate change. Ha Noi, Vietnam
MPI (2011) 5 Years socioeconomic development plant (2011–2015). Ministry of planning and investment, Ha Noi, Vietnam
MPI (2012) Vietnam socioeconomic development strategy for the period of 2011–2020. Ministry of planning and investment, Ha Noi, Vietnam
Nguyen DN, Nguyen TH (2004) Climate and Viet Nam's climate resources. Scientific and technical publishing house of Ha Noi, Vietnam
Pham NT, Phan TD (1993) Vietnam climate. Ha Noi publishing house of science and technology Ha Noi, Vietnam
Quang Ngai Department of Agriculture and Rural Development (2011a). Report on damage of crops and livestock due to extreme weather events caused. Ha Noi, Vietnam
Quang Ngai Department of Agriculture and Rural Development (2011b). Report on tested results of irrigation and river/coast erosion situation before the rainy and flood season in 2011. Ha Noi, Vietnam
Quang Ngai Department of Planning and Investment (2010) Synthesis report: overall planning of social-economic development in period 2011–2020 and Vision 2025 of Quang Ngai
Tran TX, Tran T, Hoang MT (2011) The impact of climate change on water resources Vietnam. Inst Hydrometeo Environ 300
World Bank (2010) The social dimentions of adaptation to climate change in Vietnam
Vo DP, Muhammed AB, Vo ND (2020) Climate change impact on urban flooding in Quynhon City, Vietnam